*Advance Praise for*

# BEYOND **FOSSIL FOOLS**

"Joe Shuster's *Beyond Fossil Fools* opened my eyes. It is a political, eco-nomic, and financial epiphany. The book compellingly and conversa-tionally explains why the United States must move to clean, renewable, and affordable energy in the next 30 years and how it is possible to reach that goal. The message is clear and convincing. Every policy maker, corporate executive, and thoughtful citizen must read this book and rally 'round the cause."

—Tim Penny, former Democratic U.S. congressman, 1983-1995. Co-author of *Common Cents* and *The 15 Biggest Lies in Politics*.

"As students, we hear a lot of talk, and read much about the country's energy problem and the pollution that goes with it. It's all very confusing and often contradictory; while there seems to be a big problem, nothing much, it seems, is being done about it. No one has made it understandable. That's why *Beyond Fossil Fools* is a double gift of insight and hope. It helps us see the future, encourages us to engage it, but mostly this book offers a reason to be optimistic with a ready solution—maybe the only one. Every parent, teacher and particularly every high school and college student should read this book. It is **about our future** —and how we can help shape it."

—Steven Trettel, 2006 National Junior Science Award Winner, and US representative to the 2006 International Youth Science Forum.

"*Beyond Fossil Fools* is the most factual and authoritative treatment of the energy situation that I have ever read or heard about. It does not crusade, but it does encourage us to respond to our pressing imperatives. The book is written from the perspective of a successful entrepreneur, and corporate leader on behalf of future generations. Above all, it is neither alarmist nor gloom and doom. Thankfully, it is practical and hopeful."

—Frederick M. Zimmerman
Professor Emeritus, University of St. Thomas

"Joe Shuster is a problem-solving, myth-busting, genius who illuminates the inherent problems of a world driven by *Fossil Fools* and, filled with energy foolishness. More important, Joe Shuster provides a crystal clear 30-year workable plan for providing enduring "Future Fuels" that will solve the energy problems for generations to come!

The pathway to "Energy Independence by 2040" for our children and grandchildren is precise and crystal clear, provided *Fossil Fools* readers take action today.

*Fossil Fools* is a must read and deserves to be placed on "the world's required reading list" for everyone who desires and depends on clean, affordable, and available energy for all future generations. Joe Shuster's wisdom deserves to be listened to and acted upon by all serious residents of the twenty-first Century!"

—Dr. Lyman K. Steil, CEO and Chairman of International
Listening Leadership Institute.
Author of, *Effective Listening: Key to Your Success*,
*Listening: It Can Change Your Life*, and,
*Listening Leaders: The Ten Golden Rules to Listen, Lead & Succeed.*

"I found *Beyond Fossil Fools* a down-to-earth easy to read complete book on all the important current energy issues. I'm sure this book will become a valuabale energy referrence. Shuster's conversational approach speaks to all of us, and I will make this book required reading for my 3 children. Access to affordable clean energy will clearly define their futures."

—Dan Hanlon, Educator, Author and General Manager of
Union Hill Ventures, St. Paul, MN.

"The science is accurate. The numbers are compelling. The conclusions, which we fully support, are bold. Average citizens and scientists alike are brought logically and irresistibly to the inevitable conclusion that fossil fuels cannot support us much longer... Asians, Europeans, Americans and the whole world must wrestle with the coming end of the fossil fuel era, and then adapt to a "New Economy" based on sustainable energy sources. Mr. Shuster tells us clearly and unflinchingly how to do it... This book is unique in its completeness and should be translated for all to read since it offers perhaps the only practical short-term and long-term solution to the world's most challenging problem—energy."

—Dr. Madeleine Conte, Chemical Engineer and Nuclear Scientist
—Dr. Rolland Conte, Physicist, Economist, and Author

# B E Y O N D
# FOSSIL
# FOOLS

## THE ROADMAP
## TO ENERGY
## INDEPENDENCE
## BY 2040

## JOSEPH M. SHUSTER

Beaver's Pond Press, Inc.

ISBN 10: 1-59298-235-2
ISBN 13: 978-1-59298-235-6

Library of Congress Catalog Number: 2008927002

Printed in the United States of America

First Printing: 2008

12  11  10  09  08      5  4  3  2  1

Cover and interior design by James Monroe Design, LLC.

Beaver's Pond Press, Inc.

7104 Ohms Lane, Suite 101
Edina, MN  55439
(952) 829-8818
www.BeaversPondPress.com

To order, visit www.BookHouseFulfillment.com
or call 1-800-901-3480. Reseller discounts available.

To the youth of the world so they may enjoy a reasonable future, abundant opportunities, and the blessings of freedom.

To the people and institutions that have shaped my life: my teachers, particularly Brother Timothy McNair, Tony Batistillo and Wally Miller.

To the Evans Scholarship Organization, Hibbing Junior College, and the University of Minnesota.

To my children, Susan, Peter, Siri, Pauline, and Sam, my brother Ernie, and, last but not least, Patricia, my wife of over 50 years, who was totally supportive and helped me stay focused.

# Contents

## *Part One* PROBLEMS: Fossil Foolishness

*Part Two* SOLUTIONS

## Part Three  A NEW DAWN

# CONTENTS

# List of Figures

# *Acknowledgements*

I am grateful to Kurt Burch, my editor, whose enthusiasm I appreciated and who made me look better than I am. My thanks to Jay Monroe, who created the design and layout of this book to welcome readers, and to the staff members of Beaver's Pond Press who guided me through the process of publishing this book.

My thanks also to Carol Sery, Peg Anzelc, Dean Reichow, Michael Miller, and Bill Baker for their assistance and support throughout the creation of this book.

My thanks also to the many scientists and others who generously shared their knowledge and wisdom. In particular I appreciate the cooperation and help I received from the people at Argonne National Laboratory (ANL), the National Renewable Energy Laboratory (NREL), the Natural Resources Research Institute of the University of Minnesota, and the many individuals all over the United States and from outside the country who provided information in their areas of energy expertise. William Hannum, Gerald Marsh, and George Stanford, each a retired atomic scientist from ANL, were particularly generous with their time and provided invaluable help. They deserve special recognition.

# *Preface*

Who am I and why have I written this book?

I am a successful chemical engineer, entrepreneur, and business-man, and I now see dusk setting on my career and life. I want my children and grandchildren, as well as the billions of others who live on this planet, to look into a future as bright and bountiful as the future that beckoned me for most of my life. War, poverty, and disease will probably not be eradicated. I've seen and experienced them. Yet they afflict only a portion of us at any time. However, economic and environmental disaster will likely afflict us all. That is why I foresee a set of concerns centered on energy use—resource depletion, dis-ease, and environmental decay—as the foremost problem confronting global citizens in the coming decades.

*Energy is destined to be the single most important issue of this century.* War and economic problems have always been with us, as they are today, but both of these concerns are inextricably entwined with energy matters. Energy issues will determine the kind of life future generations will live, will probably cause more wars, and, as unthink-able as it is, could become a matter of life or death for many.

I intend this book to illustrate that this serious problem is upon us. You will see an energy storm brewing—looks like the classic perfect storm—and it is closer than you think. However, I don't intend this book to be another bleak description of dire events to come. *This book proposes a quantified plan and a timetable for moving beyond fossil foolishness to a future of clean, eternal, affordable energy.* This is a book of hopeful optimism. This book is a wake-up call, a call to action, a plea. The solution solves many problems beyond affordable energy: It blunts global warming, reduces acid rain, limits ocean acidification, restricts mercury pollution, improves balance of payments, strengthens the value of the dollar, relieves transportation woes, and improves many health issues associated with fossil fuels.

Between the covers of this book you will find a broad, unblinking perspective on the energy problems of the United States and the world, including depletion and all aspects of pollution. Best of all, you will find a solution—there really is only one—and the steps to achieve the solution. The problems and solution steps are quantified and include ideas about paying for the bright, clean future. It is surprisingly affordable. There are no technological show stoppers to hinder progress. Also, if we all put our support behind the solution, I believe even the formidable political and legal barriers will fall.

The story is simple: We are running out of traditional energy resources. The world is using up available energy sources at an alarming rate. As the world burns up and burns through energy resources, energy consumption creates untold damage to people and the environment.

These two concerns—depletion and pollution—pose other problems. What alternatives to fossil fuels are available, practical, affordable, and implementable? Given reasoned projections about demand, supply, and costs, when must the transitions to such alternatives begin? Do citizens, governments, business executives, shareholders, investors, environmental advocates, bureaucrats, and other decision makers possess sufficient political will, social vision, financial resources, economic incentives, and apt institutions to spark and direct such transitions? Should governments lead the way, or should market forces set the direction and pace?

There is no silver bullet, no magic pill, no one-size-fits-all solution. We must try to dovetail many changes immediately and simultaneously.

There is another problem. The transition to new energy sources must begin immediately if the inevitable transition to alternative forms of energy is to be manageable and timely rather than chaotic, destructive, and violent.

I have no axes to grind, no vested interests to protect, no political positions to advance, no biases to hide. Neither politics nor profits motivate me. I will donate any profit from this book to energy research.

I have plowed through a small mountain of data and have consulted with countless experts. There is a lot of pure nonsense and misinformation out there. Politicians from both parties have been hostage to special interest groups resulting in poor awareness and bad energy policy. It is time that you know the truth about your country's energy problems and how they can be solved.

Although I have benefited tremendously from discussions with experts and from many publications, the opinions here are solely my own. Thank you for considering them.

I readily acknowledge that some errors and mistakes will arise in an undertaking as vast as this book. I welcome all corrections, or anything that might improve the proposed solution. I am also interested in any new, potentially viable alternative ideas. No nitpicking, please. Accompany all corrections, comments, or suggestions with supporting references, quantification, and a timeline. Address all comments to my website:

**www.beyondfossilfools.com**
I will respond to as many as possible.

# Overview

Clean, renewable, eternal energy is at hand. The energy—produced by the wind, the sun, biofuels, and nuclear power—is available, completely affordable, and fully attainable within 30 years. With this message, vision, rallying cry, and call to action, this book radiates hope, optimism, and courage. This book is a blueprint for a cleaner, safer, more peaceful, and more prosperous future.

But some will object that the world contains more than enough fossil fuel to power our future needs. Plenty of oil, plentiful natural gas, and abundant coal. I call this crowd "fossil fools." The perception of plenty is inconsistent with the reality of actual needs and dwindling reserves. If you consider current global consumption of oil and if you calculate future consumption based on worldwide population growth and economic growth, then you quickly conclude the world runs out of conventional oil reserves in about 30 years. That means the world must move to unconventional oil in oil sands, oil shale, and heavy crude oils. That's a seemingly practical and workable plan, but the likely oil still doesn't get us through this century or through your child's lifetime.

A fossil foolish objector may clear his throat and politely point out that there's no problem here because of all the coal and natural gas available to mine and tap. Yes, it is out there, but less than you think, and the world will use it quickly, especially if we must use coal to make gasoline.

But why argue over reserves and depletion dates? While there is some uncertainty, we have enough credible information and data to formulate a sound plan. And competent planning requires attention to both dwindling reserves and the other looming issue—mounting pollution. Every time fossil fuels burn, they release dangerous poisons into the environment. Finding additional fossil fuels will delay the matter of dwindling supply, but it further compounds the growing issue of deadly pollutants in our air, water, food, and bodies.

The world's twin foes—rapidly decreasing supply and rapidly accumulating pollution—are globally colossal in scope, so require an equally colossal solution. The world's task in the next 30 years is to resolve these tremendous challenges by transforming our economies, countries, cities, cultures, and perceptions away from fossil fuels toward clean, eternal, renewable energy sources.

The objector may tilt his head and clear his throat again. A pronouncement bubbles from his lips, "It seems clear to me that market forces, price increases, and entrepreneurial initiative will spark innovation and alternatives, like wind power, solar energy, and hydrogen fuel cells." Wind and solar alone or together cannot fix our energy problem. Hydrogen as a fuel is still a far-off pipe dream. Price increases can only delay the inevitable, and I for one do not want to bet the future of the world on a possible miracle. *We don't have to.*

So here's the story: Profound material and technical limits severely constrain the potential contributions of wind power and solar energy. Such limits essentially doom any contributions from hydrogen. However, breakthroughs in nuclear technologies and processes make its energy contributions clean, affordable, manageable, and sustainable. Wind, solar energy, biofuels, and other renewable sources will contribute.

That's the plan. Vigorously promote wind and solar energy so that by 2040 each contributes at least 10 percent of total U.S. electrical needs. Each would then produce more electrical energy than the total consumed in Italy. Get the remaining 80 percent from nuclear energy.

How does the United States bridge the gap from today to 2040? We succeed by simultaneously transforming the transportation fleet as rapidly as possible away from internal combustion engines to hybrid plug-ins and all-electric vehicles. At the same time, the United States and the world must aggressively develop biofuels, such as cellulosic ethanol and algae biodiesel. These transitions save huge volumes of oil and gasoline, thus stretching our reserves, limiting new pollution, and reducing demand long enough to develop and deploy required technologies and infrastructure for a wind-solar-nuclear energy system. Even so, the United States will still need unconventional oil. These transitions to energy-conserving vehicles and biofuels are the essential, irreducible, and unavoidable keys to success. Without these transformations, we run out of resources, we lack the time to move smoothly to alternatives, and we choke on the accumulating toxic pollution.

That's the story.

*Let's review:* We can and must immediately reduce our use of coal to generate electricity. We must replace coal and other fossil fuels with nuclear power and other renewable energy sources to generate electricity. However, we need time to transform transportation fleets and systems and to slash the huge volumes of petroleum our vehicles guzzle. Therefore, we need to find oil we can use for the next 30 years or longer, if need be. Thus, the United States and the world at large will need to recover oil from unconventional sources. We can partly fill our transportation needs with ethanol and biodiesel supplemented with conventionally produced oil—but that falls well short of providing enough of a bridge. That brings us to oil sands and oil shale—that is, to sources of unconventional oil.

The United States can achieve these goals by 2040 if a diligent, committed U.S. public crafts the political will. But time is running out. The world must commit time and resources *now* to develop and deploy the alternative energy sources and related infrastructures we know we will need. The U.S. public and the world at large, especially elected officials, need broader and deeper learning about energy issues. A perfect storm is brewing.

This book is arranged into three main sections: problems, solutions, and the process of getting to the new dawn.

Part I—*Fossil Foolishness*—details current problems. Chapter One, entitled "Fossil Fuels—Nature's Disappearing Act," addresses reserves and depletion. Chapter Two, "Fossil Fuel Disasters," recounts the environmental devastation caused by fossil fuels. Chapter Three, "Population," describes how population growth and accompanying industrial and economic growth greatly compound the problems of depletion and pollution. Chapter Four, called "Global Warming— The Great Distraction," challenges some of the claims made about any impending climatic apocalypse, but does not discount the possibilities. More importantly, the chapter declares that global warming, regardless of its severity and potential consequences, is only one symptom of the larger problem of fossil fuels. If the world rapidly moves away from fossil fuels, then the problems associated with human-caused global warming disappear. Chapter Five compares energy use around the world.

Part II—*Solutions*—describes contributors to an overall solution to energy problems. Chapters Six through Nine describe, in order, solar energy, wind energy, biofuels, and other renewable energy sources. Chapter Ten, entitled "Nuclear Energy—Providing a Clean, Affordable Future," discusses current nuclear technologies. Chapter Eleven, "Concerns About Nuclear Power," allays fears that many people unfamiliar with nuclear energy hold about its safety, reliability, waste, security, and potential for proliferating nuclear weapons. Chapter Twelve illustrates how nuclear and other renewable energy can also deliver a bounty of clean, fresh water, which is a pressing global need for health and agriculture. Chapter Thirteen bursts the myth of a so-called hydrogen economy as a solution to energy shortages.

Part III—*A New Dawn*—considers transitions necessary to implement clean, affordable energy and to achieve energy independence. Chapter Fourteen, "Transportation," addresses the need to replace internal combustion engines with hybrids, hybrid plug-ins, and all-electric vehicles. The chapter describes "the future is now" by detailing technological advances and describing currently available models. Chapter Fifteen, titled "Bridging the Gap," describes the need to develop unconventional oil, especially U.S. oil shale, to supplement dwindling supplies of conventional oil. Chapter Sixteen, "Energy Independence by 2040," describes a detailed plan—complete with timetables, policy requirements, funding plans, and implementation schedules—for creating a new system for generating reliable, affordable, efficient, clean, renewable energy that resolves the twin dilemmas of fossil fuels (depletion and pollution) and that achieves energy independence.

## HOW ABOUT IT? ENERGY INDEPENDENCE DAY, July 4, 2040?

# Part One

## PROBLEMS
FOSSIL FOOLISHNESS

# *Chapter 1*

## FOSSIL FUELS

### NATURE'S DISAPPEARING ACT

The world has had an incredible economic ride on the back of fossil fuels—coal, natural gas, and oil. We didn't know there would be serious hidden costs to pay for using fossil fuels. Once we started, there was no turning back. Fossil fuels realigned the world, caused wars, and now confront the world with two colossal problems requiring equally colossal solutions. One, the world is running out of convenient, easy-to-find fossil fuels, and we are running out at an accelerating rate. (Please see Figure 1.1, the Countdown to Total Depletion, on the inside front cover. This chart is the foundation of this book, a compact picture of the energy problems we face.) Two, as we keep burning coal, oil, gasoline, and natural gas, we risk irreversible, devastating damage to our common home, the Earth.

In a little more than a century we have used up one-half of the world's conventional oil reserves and almost half of the world's natural gas reserves. Use is rapidly increasing with population growth and the emergence of China and India as economic powerhouses.

Fossil fuels supply most of the energy consumed in the world

9

today. They are concentrated in the ground and easy to recover, so are a cheap energy source, or so we thought. They are not cheap however when you count the terribly high costs of pollution and the damage to the environment, animals, and people. While we pay for electricity only pennies per kilowatt hour, the real cost of the damage to our planet is incalculable.

Second, we are poisoning ourselves. The pollution caused by the burning of fossil fuels amounts to slow, global suicide. Mass suicide by contaminants? Genocide by greenhouse gases? Even the words sound frightening: toxins, carcinogens, pollutants, effluents, particulates. As use increases, the tons of pollution pile up, blow around, circulate, settle, and end up in our food, lungs, and blood. And even if you don't believe some claims about global warming, even the harshest critics acknowledge that the burning of fossil fuels contributes *something* to climate change.

## FOSSIL FUELS in the UNITED STATES

### Energy Sources and Flows in the United States

Take a look at Figure 1.2. It shows where the energy the United States used in 2002 came from and how we used it, from input to output. The overall pattern is the same today. Pay attention to the relative values. The energy unit used, the quad, is not my main point. But for the record, one *quad* equals one quadrillion British Thermal Units (BTU). That's 1,000,000,000,000,000 BTU. That equals 293 billion kilowatt hours.

Figure 1.2 may look confusing, but it isn't. Think of the Figure as illustrating the major energy "rivers" and significant "tributaries" carrying the flows of energy in the United States. The "sources" are labeled at the far-left side, and the "destinations" or "uses" appear on the right side. Notice how much more energy is lost than is used usefully.

Information from Figure 1.2 produced Figure 1.3, which lists U.S. energy use in percentages, not quads. Note that 21 percent of the

# Figure 1.2. U.S. Energy Flows Measured in Quads, 2002; Net Primary Resource Consumption, Totals about 97 Quads of Energy

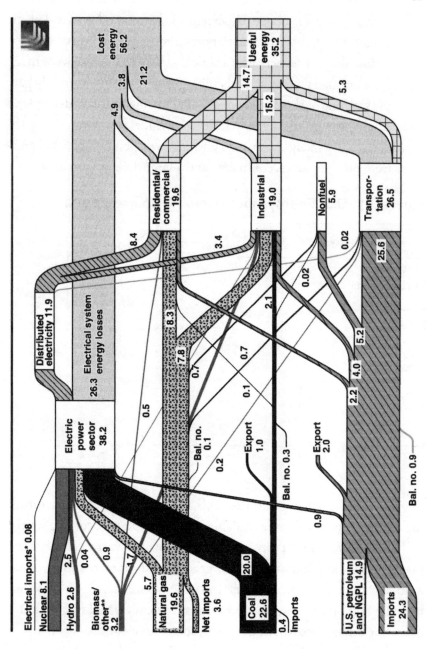

Sources: Lawrence Livermore National Laboratory (2003, 2004) at <http://eed.llnl. gov/flow> and Energy Information Administration, *Annual Energy Review 2002.*

electricity generated in the United States comes from nuclear reactors. Also note the disgusting, embarrassing, and irresponsible fact that the United States imports about 62 percent of the total oil it uses. Let me say it again: The United States imports almost two-thirds of all oil it burns. How did our leaders ever think this was ok? There were many energy-danger signals, and the oil embargos of the 1970s certainly gave us ample warning of the consequences of not being energy independent. Needed changes should already have been underway. Don't we wish? Our leaders most likely chose to do nothing in response to political pressures from oil companies, confusion, and a large dose of ignorance.

## Figure 1.3. U.S. Energy Inputs from All Sources, 2002

| Energy Source | Percent of Total Energy Generated | Percent of Total Electricity Generated |
|---|---|---|
| NUCLEAR | 8.2 | 21.3 |
| HYDROPOWER (DAMS) | 2.6 | 6.6 |
| BIOMASS AND OTHER | 3.3 | 2.4 |
| COAL | 23.2 | 52.5 |
| Domestic | 22.8 | -- |
| Imported | 0.4 | -- |
| NATURAL GAS | 23.3 | 14.9 |
| Domestic | 19.7 | -- |
| Imported | 3.6 | -- |
| OIL | 39.4 | 2.3 |
| Domestic | 15.0 | -- |
| Imported | 24.4 | -- |
| TOTALS | 100.0 | 100.0 |

Biomass, according to the U.S. Department of Agriculture, includes any organic matter available as a renewable or recurring source, including agricultural crops, trees, wood, wood waste and residues, plants (including aquatic plants), grasses and grass residues, fibers, animal wastes, municipal wastes, and other waste materials. Together, biomass and other energy sources not specifically listed in Figure 1.3, including ethanol, geothermal energy, solar energy, and wind energy— accounted for only 3.3 percent of total energy produced in the United States in 2002. Over 3 percent is biomass alone. Geothermal (0.3

percent), solar (0.06 percent), and wind (0.1 percent) *combined* account for much less than 1 percent—an inconsequential 1 percent. As you read about the dramatic growth of wind and solar power, be aware that this is growth from a very low base. Everything helps, but the world's mammoth energy problems require mammoth solutions. So far, most politicians, including Al Gore, have recommended ridiculously anemic Band-Aid solutions. As they seduced us with their non-solutions, confusion and inaction resulted.

The following sections look in detail at coal, natural gas, and oil. These fuels took hundreds of millions of years to develop from green matter such as algae, grasses, and other plants. All green mass on earth grew from the process of photosynthesis, probably the most important chemical reaction on earth. All life depends on it. Plants grow by taking carbon dioxide from the atmosphere. Photosynthesis uses carbon dioxide, water, and sunlight to make glucose (sugar) and the oxygen we breathe. Since carbon dioxide sustains all life on earth, it is not a pollutant. Rather, carbon dioxide makes the world habitable by helping to regulate the earth's temperature in a range suitable for plant and animal life. Carbon dioxide, like many things, becomes a pollutant only when in excess.

> All green mass on earth grew from the process of photosynthesis, probably the most important chemical reaction on earth. All life depends on it.

I start with coal because people have burned coal for heat for thousands of years. In contrast, only since World War II has natural gas become a very significant energy source. Oil and petroleum products have been the dominant sources for generating energy only since the nineteenth century. For example, John D. Rockefeller founded Standard Oil in 1881.

## COAL

Coal fueled the Industrial Revolution in the eighteenth and nineteenth centuries. It produced the steam that drove factory machines, steamboats, and trains. Today coal is used primarily to generate electricity. While new technologies help coal burn more cleanly, it remains the most polluting energy source. Again, please refer to the Depletion Chart (Figure 1.1) for information on reserves and consumption. The world will run out of coal in less than 200 years. If the world needs to convert coal to oil, then it would last less than 100 years.

### Coal and Electricity

Coal is a robust energy source and a tireless workhorse for generating electricity. About 90 percent of all mined coal is used to generate electricity. In the United States more than one-half of all generated electricity comes from burning coal.

The world's recoverable coal reserves (see Figure 1.4) are estimated at 1 trillion short tons (2,000 pounds per ton), a large and meaningless number unless put in some context. Coal contains an average of 20 million BTU per ton. One ton of coal contains about the same energy as 160 gallons of gasoline.

Coal comes in different grades and degrees of quality. As green plants decompose, they eventually become usable fuel called peat or peat moss. After further decay and pressure, the peat turns into lignite (coal). Further chemical changes, combined with continued pressure that drives out water and increases the carbon content, creates sub-bituminous, bituminous, and anthracite coals. For comparison: One pound of lignite has a heating value of approximately 6000 BTU; bituminous has about 15,000 BTU per pound. One pound of lignite can produce enough usable electrical energy to light a 100-watt light bulb for approximately 5 hours, bituminous for 13 hours, and anthracite for 17 hours.

**Figure 1.4. Percentage of World's Recoverable Coal Reserves, 2005**

| | |
|---|---|
| United States | 27 |
| Russia | 17 |
| European Union | 04 |
| India | 10 |
| China | 13 |
| Rest of World | 29 |
| TOTAL | 100% |

The burning of coal and fossil fuels is terribly inefficient. Only about one-third of the potential energy of coal actually generates electricity. The remaining two-thirds is lost or rejected through inefficiencies and heat. It goes up in smoke, right up the stack. For comparison, in the United States the overall efficiency of all used fuels, including nuclear, is about 36 percent. A mere 36 percent. The production of electricity is about 31 percent efficient. The burning of fuels for transportation is about 20 percent efficient. Generating energy for residential and commercial use is about 38 percent efficient. It's 57 percent for industrial uses.

> Only about one-third of the potential energy of coal actually generates electricity. The remaining two-thirds is lost or rejected through inefficiencies and heat. It goes up in smoke, right up the stack.

Surface mining produces most of the coal in the United States. *Surface coal* usually sits less than 200 feet below the surface. Yet some mines burrow 1000 feet underground. We often read about coal mining accidents deep underground, which seem to be chronic worldwide. The U.S. coal industry has done a pretty good job of reducing the negative environmental effects of mining. The industry restores disturbed lands and goes to great lengths to prevent damage to rivers, streams, lakes, ponds, and underground waters.

## Coal and Pollution

Coal produces a lot of stuff you don't want to eat or breathe. Although coal served the world well in the past, we cannot continue to burn coal because it produces the most and the foulest pollution, creating harm and damage that should be unacceptable to any society. At the mines, the coal is processed to remove or reduce dirt, rock, ash, sulfur, and other unwanted materials. This processing increases the heating value of the coal and reduces pollutants. Cleaning up smokestacks helps, but doesn't do enough. Burning cleaner anthracite coal helps, but there isn't enough of it. Coal gasification—turning coal into gas—helps, but the process is inefficient, expensive, and still pollutes.

Consider these foul facts about coal and carbon dioxide ($CO_2$). Coal-fired power plants emit approximately 36 percent of the $CO_2$ in the United States. Here's the simple story: Burning 1 pound of coal generates about 3.4 pounds of $CO_2$—that is, 3.4 tons of $CO_2$ for each ton of coal burned. And that's just the carbon dioxide. Burning coal also produces other gases that cause acid rain, smog, and the acidification of the oceans. Mercury emissions are another problem.

Burning 1 pound of coal generates about 3.4 pounds of $CO_2$—that is, 3.4 tons of $CO_2$ for each ton of coal burned.

Bottom line: Stop burning coal—period. Build no new coal-fired plants, and start decommissioning existing plants. We can do it, and the United States and the world can easily afford it.

## NATURAL GAS

Natural gas, like underground petroleum, is millions of years old. In ancient times, natural gas was the root of myth and superstition, because when lightning struck a seepage, fire instantly burst. One of these flares, believed by ancient Greeks to be of divine origin, was discovered by a goat herdsman in 1000 BC and became the site of a famous temple. The temple housed a priestess, who became famous as the Oracle of Delphi, who issued many prophesies said to be inspired by the flame.

In 1785 Britain was the first country to use natural gas for lighting. In 1816 Baltimore, Maryland, used natural gas to light its streets. In 1895 Robert Bunsen invented what became known as the Bunsen Burner, which was subsequently used for cooking, heating, and laboratory experiments. However, prior to World War II, natural gas was most often vented into the atmosphere or flared when found alongside oil. It was a nuisance. Since World War II, the industry grew rapidly.

Natural gas is the cleanest burning fossil fuel. Each pound of natural gas burned generates 2.75 pounds of carbon dioxide. While natural gas produces less $CO_2$ than coal or oil, it still creates millions of tons of carbon dioxide in the atmosphere each year. While often called the "clean fuel," natural gas is no great bargain when it comes to $CO_2$, producing only a little less $CO_2$ pollution than coal does. Natural gas (methane) is itself a powerful greenhouse gas, about 22 times worse than $CO_2$.

At present consumption rates, the world will run out of natural gas in less than 65 years. If we add population growth, then we run out in 50 years. Between 2020 and 2030 global use of natural gas will exceed production or new finds, after which demand will be greater than the discovery of new reserves. Reserves will then decline at an accelerating rate. U.S. reserves of natural gas will peak in the coming decade. Three-fourths of the world's supply of natural gas sits in the Middle East and Russia.

So why are we switching to natural gas at the very time its production will begin to lag further and further behind use? We switch because we have few options in the near future. To keep up with future demand, we must look desperately for new sources of natural gas. That won't be easy since the most promising sites have already been found. Is the United States willing to bet that we'll hit the jackpot, win the natural-gas lottery,

Meeting anticipated demand for natural gas in 2015 would cost approximately $800 billion.

or find the buried treasure that allows us to meet current and future demand? I don't want to sit around waiting for luck or good fortune to find us. I am a gambler, but this gamble would, by any measure, be asinine. Even if we hit the natural gas mother lode, its use would still pollute unacceptably—particularly if you believe $CO_2$ is the major cause of global warming.

A 1999 study by the National Petroleum Council projected that meeting anticipated demand for natural gas in the United States in 2015—only 7 years from now as I write this—would cost approximately $800 billion in drilling, supply development, pipeline construction, and other costs. That's a huge figure, and the money is better spent on other projects, as I'll illustrate. The nation should not make this massive investment in a losing cause. In spite of projections of new discoveries, natural gas prices will vary wildly, but the prices will continue to rise. Supply and demand dictates that prices will escalate dramatically in a herky-jerky, undisciplined manner, but always up, up, and up.

So the cleanest fossil fuel is neither clean enough nor plentiful enough. Yet since natural gas produces fewer pollutants than coal, oil, and gasoline, it must be used during the 30-year transition to all-renewable, eternal energy sources.

## OIL, PETROLEUM, and GASOLINE

Oil, the indispensable fossil fuel, is the only source for producing all-important portable, liquid transportation fuels. A barrel of oil is also the source of other fuels and products (see Figure 1.5). I wrote a report on the energy crisis for congressional leaders in 1973. At that time the United States had 100 billion barrels of oil in reserve. Now U.S. reserves have only 22 billion barrels. U.S. oil reserves peaked in 1997. The world's reserves of conventional oil will peak sometime in the next decade. In 2005, the United States used more than 20 million barrels of oil per day, approximately 7.5 billion barrels of oil per year. About 3 billion barrels of

## Figure 1.5. Products Made from a Barrel of Crude Oil

PRODUCTS MADE FROM
A BARREL OF CRUDE OIL

GASOLINE: 19.7

DIESEL: 10

JET FUEL: 4

HEAVY OIL: 1.8

OTHER: 6.5

TOTAL: 42 Gallons

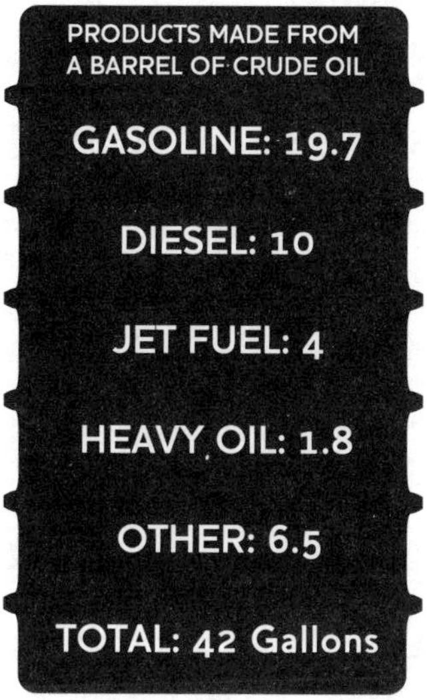

the oil the United States consumes annually comes from U.S. oil reserves; imports make up the remaining 4.5 billion barrels per year. You do the math. If the United States were forced to use only its own oil, then the nation would be *totally out* in approximately 3 years. Three years to bone dry. So the United States *must* secure and import oil. Unavoidable, no matter the cost in dollars and lives. Does that help you understand the U.S. military in Iraq? (For details on the sources of U.S. oil imports, see Figure 1.6.)

The U.S. peak in oil production is already in our rear-view mirror—behind us. The only possible relief is to move immediately to clean, non-polluting renewable energy and to change the kinds of cars and trucks we drive.

Perhaps now you begin to see the future collision of trends in the next 30 years—the "perfect storm" for an energy crisis is gathering and brewing. Some see this storm coming. Some don't or won't see, preferring denial or ignorance. Yet the storm is there. The United States and the world are literally running out of gas, oil, and even coal. Add population growth and increasing demand, and the storm gathers fury. Even an observant idiot can see the darkening horizons and the awaiting disaster. What does this tell us about our fearless leaders? Idiots? Fossil fools? Not really, just uninformed, misinformed, or beholden to special interest groups.

### Figure 1.6. Top Ten Sources of U.S. Petroleum Imports, September 2007 (Thousands of Barrels Per Day)

| Country | Barrels | Approximate % of U.S. Oil Imports |
|---|---|---|
| Canada | 2,467 | 20 |
| Saudi Arabia | 1,560 | 13 |
| Mexico | 1,429 | 12 |
| Venezuela | 1,325 | 11 |
| Nigeria | 1,181 | 10 |
| Algeria | 701 | 6 |
| Iraq | 603 | 5 |
| Angola | 591 | 5 |
| Virgin Islands | 381 | 3 |
| Russia | 348 | 3 |
| Rest of World | 1,400 | 12 |

Source: Energy Information Administration at <http://www.eia.doe.gov/pub/oil_gas/petroleum/data_publications/company_level_imports/current/import.html>.

Even over the next 10–20 years there will be little reduction in U.S. oil imports. Measured by 2007 consumption rates, the world will run out of conventional oil in less than 40 years. With predicted population increases and business-as-usual practices, the world will run out of conventional oil in less than 30 years. For the sake of argument, let's say the world taps 1 trillion barrels of oil from the huge potential global reserves of unconventional sources, like oil sands and oil shale. That only doubles the timeline. The perfect global storm will strike well within the expected lifetime of an infant. Why bother to plan for personal retirement, but not plan for the "retirement" and demise of fossil fuels?—your retirement won't be worth much if you don't.

If the imports and dependence don't scare you, then it will certainly terrify your children and grandchildren—that is, it would scare them if they had any idea about what is going on. They will likely have to fight for oil. Some of our sons and daughters and grandkids are already fighting on battlefields in the Middle East.

Even over the next 10–20 years there will be little reduction in U.S. oil imports.

Moreover, as we run out, the next generation could be deprived the use of oil as a chemical feed stock to make plastics and other products.

The numbers clearly and unmistakably tell us that we must begin today to shift to other fuel sources and abandon the internal combustion engine. This transition is absolutely necessary, inevitable, and unavoidable.

This transition is no arid, dull, technical matter. Fossil fuels have been a huge element in the foundation of contemporary civilization— economic activity, transportation, communication, lifestyles, the works. However, the transition to another source of energy, while doable and affordable, will require a fundamental overhaul in the physical infrastructure and basic ideas and assumptions of society.

## Consumption and Reserves

Of course, the world will never absolutely run out of oil. The price will continue to rise until the price is beyond the means of most people and firms. They will turn to other fuel sources or to "non-market" means for acquiring what they need. "Non-market" is a code-word for war, violence, and crime.

Yes, it is true that crises tend to reduce consumers' demand for gasoline. For example, the oil embargos of 1973 and 1979 and the terrorist attack on the World Trade Center in 2001 decreased demand. Yet it is also true that demand quickly returns to previous levels, especially as economies regain strength. Any trouble in the Middle East seems to cause a spike in oil and gasoline prices. We grumble, we gripe, and we pay. Don't be fooled by any short-term easing of oil prices after price spikes. The general trend is unmistakable: up, up, and away. Prices for oil and gasoline continue to rise in response to our ever-expanding economies and our relentlessly growing global population. These demands for energy resources will cause oil prices to rise steadily, then soar. Then the oil will be gone, or effectively unavailable, because it is too expensive.

Not all so-called experts agree on the date when world oil reserves will peak. If not this decade, then the next one. Soon, by any estimate.

Some argue that advances in technology will make factories, machines, and vehicles more energy-efficient, so will push the peak-oil date further into the future, even to 2050 and beyond. It won't. Conservation can buy some time, but only 10 years at the most. Buying us a few years or a decade is important, but is otherwise meaningless unless we take advantage of the time we buy. The big picture remains clear and ominous: The world is rapidly depleting conventional energy sources, and this depletion is made worse by growing populations, increasing demand, and the simultaneous creation of pollution.

> The big picture remains clear and ominous: The world is rapidly depleting conventional energy sources, and this depletion is made worse by growing populations, increasing demand, and the simultaneous creation of pollution.

Even as the world turns to unconventional sources, such as oil sands and oil shale, the pollution problem marches on. The issue isn't whether we run out, or even when we run out. The issue is simpler: We know it will run out and we know it will run out soon. We must begin the retreat from fossil fuels now.

Optimistic projections continue to distract us. Mr. Hofmeister, president of Shell Oil Company, made a 50-city tour to assure audiences that we will be okay if firms drill offshore, drill in Alaska, and so on. He was simply coughing up the industry's biased views. He offered so much spin, I became dizzy. I asked him how he got his numbers and conclusions, because our data didn't agree. He said not to worry because we'd simply find more oil when we need it. I wonder if he believed what he was saying. After the meeting, one of his lieutenants asked, "Just who do you represent?" I replied, "My grandchildren."

> U.S. conventional oil reserves will last only 3 years at our present rate of consumption, if we stop importing oil.

We cannot afford delay and inaction, particularly since we have solutions at hand, but I fear we may be too late. We've got to begin *now* to build large-scale facilities for generating energy from biofuels, wind, solar, and nuclear sources—and we must do it in 30 years. But let's look at our depletion problems without yet considering the twin problem of pollution. Study the main Depletion Chart (Figure 1.1)—it tells the whole story.

## Depletion

How much petroleum is left in the ground? Seems a simple enough question, but answers vary, often by a lot. For example, some members of the Organization of Petroleum Exporting Countries (OPEC) likely inflate the figures describing their reserves, because their "official reserves" influence the volume of oil they are allowed to pump and sell.

> At the world's current rate of use of oil—30 billion barrels per year—conventional oil reserves will last 37 years. By 2044 the world is bone dry.

Again, the importance of the Depletion Figures cannot be overstated. Figure 1.1 was prepared from data obtained from several of the most credible sources available: Central Intelligence Agency (US-CIA), Department of Energy (US-DOE), Energy Information Administration (US-EIA), and the World Energy Council (WEC). Figures from official U.S. government sources are widely recognized as the most accurate. However, even these numbers often vary. Different sources calculate different figures with different assumptions. Although some figures may be off somewhat, I am confident that the overall numbers and projections are quite accurate, because they are a composite of many credible sources.

### When do we run out?

At current consumption rates, even doubling the world's *total oil* with 1.1 trillion barrels from unconventional sources makes global oil reserves last only 74 years to bone dry. That's just to 2082, well within the life expectancies of any child born in the United States this year. Now let's modestly complicate this simple math by adding the effects of population and economic growth. With increased consumption from population and economic growth, the conventional oil would last only 28 more years (to about 2035), and when doubling the oil reserves by speculatively adding unconventional sources, the oil would last only about 55 more years (to about 2063).

The perfect storm gathers strength.

### Don't forget China and India

Let's also look at this from another perspective. Let's say that by 2037 the demand for oil in China and India increases to just 4.4 and 1.8 barrels per person per year, respectively. On a per capita basis, China would then consume about 17 percent, and India would consume only about 7 percent, of the per capita use of the United States. While the increase in these countries is modest, it would dramatically affect the world's total oil consumption. By adding the predicted population increase of 900 million for these countries in the next 30 years, we identify another vector of the impending perfect storm.

It doesn't matter how you count and factor the data. The bottom line is clear. The world is quickly running out of cheap, available oil, natural gas, and even coal, especially if the United States and the world continue their bad energy habits. Experts predict that the consumption of oil will increase by 1–2 percent per year over the next several decades, based on growth of the global population over the same decades and on the accelerating economic growth of China and India.

> The world is quickly running out of cheap, available oil, natural gas, and even coal, especially if the United States and the world continue their bad energy habits.

## Conservation

Advocates and talking heads have been telling the public for a long time that conservation is the answer. This was a large part of Al Gore's erroneous conclusions in *An Inconvenient Truth*. It is important, therefore, that we all understand what conservation can do for us.

Let's see what happens if the world conserves 10 percent of the energy it uses. At present consumption rates, the world's oil would last 4 more years, natural gas 7 more years, and coal 18 more years. If the world conserves a full 20 percent, then these numbers will double: 8, 14, and 36 years respectively. To conserve 20 percent would be very difficult and would not improve the overall picture very much, except to buy us some time for the transition to renewable energy. And even this would be meaningless unless we take advantage of the time saved. Surely a worthy effort, but do not swallow the idea that people in the United States or the world can conserve themselves out of the problem.

If the United States conserves 10 percent, then oil would last 4 months longer, natural gas would last an additional year, and coal would last 25 extra years.

## SOME CONCLUSIONS, SIMPLE TRUTHS, and CONFUSIONS

Policymakers and so-called experts often make predictions and policies based on inaccurate, confusing, or confused data wrapped around their own biases. National leaders can be led to policy conclusions based on faulty or false data. Lobbyists add another layer of skewed, biased data. The result is manufactured "facts" that support a desired view, rather than views being formed around real facts. Terrible policies get created in just this way.

Consider the 2005 U.S. energy bill. Can anybody explain to the citizens of the United States how this bill solves the country's energy problem in time? The politicians rejoiced, shouted, and cheered. They told us they wrestled with the energy problem and won. They thumped their collective chests and told us there would be no crisis if we just let them drill for more oil. Did anybody actually do the math? Did anybody actually calculate consumption rates in relation to reserves?

It was mostly a business-as-usual bill supporting the oil, natural gas, and coal industries. To its credit the bill provided some support for nuclear energy, biomass, and other renewable sources. While parts of the bill were directed toward a cleaner environment, there was nothing that could possibly bring the nation much closer to energy independence. The parochial bill did not acknowledge the global aspects of the problem. Any U.S. energy bill that is not related to the world's energy problems makes little sense.

The Energy Independence and Security Act of 2007 is not much better. Biofuels get a welcome boost and research is expanded for solar, geothermal, and batteries, which makes a lot of sense. It expands the U.S. Department of Energy's program for capturing and sequestering carbon dioxide, which will waste a lot of time. The program is intended to expand the use of coal—not a good thing. Most of the bill is directed to conservation, which sets new fuel standards for autos and light trucks—Band-Aid, feel-good fixes that buy some time but will not prevent future economic disaster. Overall, the time lines are too long and the bill will do very little to take the country to energy independence or provide much security. We must insist on our government showing the American people how this legislation takes us to energy independence. *It can't.* We need verifiable quantification with a definite timeline. Indeed we need a new bill in tune with U.S. and world energy realities.

If policymakers can't save us, then Alaskan oil and new technologies will, right?

## Alaskan Oil Will Save Us?

Does the public benefit from drilling in environmentally sensitive areas? Or do oil companies benefit the most?

U.S. drivers would blow the entire ANWR reserves out of their collective tailpipes in less than 8 months.

Most sources say the United States has about 20–23 billion barrels of oil in reserves, down from 100 billion barrels in 1973. Let's say that drilling in the Arctic National Wildlife Refuge (ANWR) yields 5 billion additional barrels, as reported by Joel K. Bourne in *National Geographic*. For perspective, 5 billion barrels would be more than the present total oil reserves in Alaska (which total approximately 4 billion barrels) and only about

20 percent of all U.S. reserves. If priced at $80 per barrel, then 5 billion barrels in ANWR would be worth $400 billion. This is a lot of money, and the oil seems like a big deal, but is it? U.S. drivers would blow the entire ANWR reserves out of their collective tailpipes in less than 8 months. And we haven't discussed the obvious environmental damage to ANWR.

Here's the simple truth: Drilling in ANWR, the National Petroleum Reserve in Alaska (NPRA), and the Beaufort Sea (that portion of the Arctic Ocean north of Alaska and of the Canadian territory of Yukon) is at best another Band-Aid solution. Recovering oil from these areas offers little relief and poses grave environmental harm. Drilling in the Beaufort Sea threatens waters teaming with plankton and krill, fundamental foods of life at the beginning of the food chain. A spill in the Beaufort Sea could result in a tragedy of epic proportions, a disaster that could make the 1989 spill of the *Exxon Valdez*, one of the world's largest man-made environmental disasters, look trivial. Alaskan oil should remain in the ground—period. It solves nothing. It only adds to the bottom line of some oil companies and their shareholders.

## Technology Will Solve the Problem?

Several technological advances could extend or increase oil reserves. These advances include deep-ocean drilling, directional drilling, seismic analysis, new injection technologies, and recovering oil from oil sands and oil shale.

*Deep-ocean drilling* means going after oil about one-half mile or more beneath the ocean surface. Of course, there is always the possibility of a devastating oil spill or leak. However, deep-ocean drilling becomes more attractive as the oil industry better learns how to find this oil. Analysts project that this expensive process will increase global oil reserves by about 5 percent. Great, but the world would gobble up this oil and push it out our exhaust pipes in less than 2 years. But what if only the United States were to consume this oil? Then it would last a paltry 7 years—really not enough to make a difference. Seven years is shorter than the life of most cars.

*Directional drilling*, a less expensive technique, more accurately guides the drill to where the oil is located. *Advanced seismic techniques* also better pinpoint where oil is located. These techniques can increase yields by 15 percent in some fields, but they also increase costs by about the same amount—a good buy as scarcity increases.

*Injection methods* "inject" water, natural gas, steam, nitrogen, or carbon dioxide into the drill shafts to "flush out" remaining oil. This technique increases the yield, but also increases costs by 50 percent or more. Also a good buy.

Recovering oil from *oil sands*, a complex and messy process, is economical at the present price of oil. The cost is about $20–25 per barrel. Alberta, Canada, has huge deposits of oil sands, which are mined using power shovels, then transported by huge trucks to processing plants. The costs for such plants are high, and utilities and investors are said to be reluctant to build plants for fear of oil prices falling dramatically or steadily to below production costs. Not a chance. Yet oil sands and oil shale are indispensable for the 30-year transition to clean, renewable energy sources.

The United States has huge deposits of *oil shale*. Actual recoverable oil could be from hundreds of billions of barrels to trillions of barrels. The costs of recovering oil from oil shale varies with the concentration of oil, but will be somewhat more expensive than recovering oil from oil sands. Although estimates of cost vary greatly, I believe that a mature oil shale industry will be able to produce oil for $25–40 per barrel and could conceivably fall below $20 per barrel for the more concentrated deposits.

In short, the first 4 beneficial techniques do not provide much comfort since they extend oil supplies only a decade at most. Recovering oil from oil sands and oil shale could help "bridge" the United States and the world to energy independence and renewable energy sources in 30 years, although we would continue to pay a high price in pollution. (See Chapter Fifteen for extended discussions of Canadian oil sands and U.S. oil shale.) Of course, the faster U.S. firms and citizens convert to all-renewable sources, the less polluting oil we will need from any source to bridge us. Conventional oil just won't get the United States or the world to energy independence unless the transformation takes less than 20 years. Since we all know such a 20-year transformation won't happen for a number of reasons we can easily anticipate, we also learn how desperate the situation really is. With business as usual, it would take at least a decade to build the infrastructure to recover oil from oil shale. Here again the situation is so pressing that we must move to a crash program.

Let's turn now to the costs of fossil fuels.

Isn't cancer caused by
fossil fuel a cost?

## THE REAL COSTS of FOSSIL FUELS

Fossil fuels appear "cheap" and efficient—or so the public has been led to believe. But fossil fuels become a lot more costly if you calculate the terribly high costs of pollution and damage to the environment, animals, and people. Most people don't include the costs of radiation treatments and oxygen tents in their costs for gasoline, diesel fuel, heating oil, and electricity. But isn't cancer caused by fossil fuel a cost? Isn't contaminated food a cost? Isn't acid rain a cost? Isn't a mammoth defense budget directed in part at protecting oil flows a cost? One prominent source estimates that the cost of "securing access" to Middle Eastern oil costs the U.S. government $50 billion per year, *independent of the $12 billion per month it costs to wage war in Iraq and Afghanistan* [Institute for Analysis of Global Security, http://www. iags.org/costofoil.html]. And the United States imports only about 12 percent of its oil from the Middle East, where about 65 percent of global oil reserves are located.

If we start calculating costs this way—that is, if we calculate *social costs*—then it becomes clear that the price of "cheap energy" could cost us the future health of our planet, our economies, and our children. While we pay only 7¢–12¢ per kilowatt hour for electricity, the *real cost* of electricity to our planet, our health, and our prosperity is incalculable.

No matter what we do, we cannot get away from the high cost to the world of pollution caused by using fossil fuels. We can reduce such costs, but we cannot avoid them. Pollutants will *always* be released when burning fossil fuels—it's basic chemistry. Nobody can change it. The fewer the pollutants we want, the greater the costs we'll pay. The only certain way to avoid producing pollution and harming health and the environment is to quit burning all fossil fuels. Anything else is fossil foolishness. For more on fossil fuel disasters and costs, see Chapter Two. For information on global warming, see Chapter Four.

## BOTTOM LINE

During the next 30 years the world's citizens and governments will begin feeling serious pain as they wrestle with increasingly grim aspects of the energy crisis, such as conflicts over resources and acute shortages of water, food, and fuel. Even if we double existing oil reserves, then the crisis still hits well within the lifetimes of most people alive today. If the "energy problem" sounds too abstract, then look into the eyes of a preschool child. Explain to him or her why nothing is being done and why life's opportunities, as we know them, may not be there in the future.

I don't want my children and grandchildren to face a bleak world of depleted fossil fuels and the pollution they cause. How about you?

- A transition to clean, affordable, renewable energy will be surprisingly affordable, but we must act now with all the resources we can bring to bear. The worst thing is to do nothing or delay. We no longer have the gift of time.

- Why on earth would we spend a staggering amount of money to keep fighting a battle we cannot win? We can't defeat simple chemistry, and we can't defeat depletion rates. We must *abandon fossil fuels*, not look for ways to extend their lives.

How? Read on.

# Chapter 2

## FOSSIL FUEL DISASTERS

The *depletion* of fossil fuels is a serious global and national problem, but that's not all. Fossil fuels cause *pollution disasters*—there is no other term for it.

Although coal-burning power plants are by far the worst polluters in the United States and the world, the burning of fossil fuels from any source creates prodigious amounts of pollutants, including carbon dioxide ($CO_2$), sulfur dioxide ($SO_2$), nitrous oxide ($N_2O$), and heavy metals, such as mercury. These impurities and carcinogens in coal go right up the smokestacks and down our throats into our lungs and bellies.

The burning of natural gas (methane) illustrates how burning a fossil fuel creates environmental and health disasters. Methane in combination with oxygen produces carbon dioxide, water, and heat when burned. This reaction tells us the precise amount of $CO_2$ produced. Get this—for every *pound*

For every pound of methane we burn, we also produce 2.75 pounds of $CO_2$. For every gallon of gasoline we burn, we produce about 17 pounds of $CO_2$.

of methane we burn, we also produce 2.75 pounds of $CO_2$. For every *gallon* of gasoline we burn, we produce about 17 pounds of $CO_2$. If you get only 17 miles per gallon, and many of you don't do any better, then you produce about 1 pound of $CO_2$ for every mile you drive. If you drive 15,000 miles per year, you pump 7.5 tons of $CO_2$ out of your tailpipe annually. Now multiply this by the 250 million cars and light trucks on U.S. roads.

And let's not forget global warming. Global warming is such a "hot" topic, that I discuss it in a separate chapter. It is important to understand that global warming—no matter what kind of problem you think it is—is a mere symptom of the burning of fossil fuels, a symptom just like depletion, acid rain, ocean acidification, mercury poisoning, and the like. The main problem is the burning of fossil fuels, not global warming.

**More people are killed by air pollution than by auto accidents.**

Reducing the use of fossil fuels is not good enough to prevent further destruction of the environment and human health. Any use means depletion and pollution. Unless the world abandons fossil fuels completely, we will continue killing ourselves and others with pollution, particularly as out-of-control population growth drives demand and the use of fossil fuels. Further, the burning of fossil fuels prematurely kills people from asthma, heart disease, and lung disorders. Luis Cifuentes and other researchers at Carnegie Mellon University in Pittsburgh studied the health effects of fossil-fuel pollution on death rates in 4 international cities: São Paulo, Brazil; Mexico City, Mexico; Santiago, Chile; and New York City, USA. The researchers estimate that *more people are killed by air pollution than by auto accidents.* Now *that* is a tangible fossil-fuel disaster.

Many countries established "quality standards" for air and water. The public must encourage such standards, but they represent a naïve long-term policy. A finger in a dike. Standards improve conditions by addressing symptoms, not the underlying disease. We simply cannot change the laws of chemistry—for every molecule of carbon, sulfur, and nitrogen we burn, we create toxic pollution.

*Coal*-fired power plants emit approximately 36 percent of U.S. carbon dioxide emissions, more than 60 percent of the sulfur dioxide, 23 percent of the nitrous oxide, and the majority of mercury poison. The burning of *oil and gasoline* contribute large amounts of carbon dioxide and other greenhouse gases. Pound for pound, the burning of *natural gas* produces almost 90 percent of the carbon dioxide produced by burning gasoline and 80 percent of the carbon dioxide produced by burning coal. So the allegedly "clean" fossil fuel is not so clean after all.

Pollutants are either directly released into the environment, or they are formed by subsequent chemical reactions. A *direct-release* pollutant—such as carbon dioxide, carbon monoxide, sulfur dioxide, nitrous oxide, and some metal vapors—is emitted directly from a given source and produced by burning coal and gasoline. A *subsequent* pollutant is formed through a chemical reaction involving direct-release pollutants. The formation of acidic oceans, acid rain, and photochemical ozone/smog are important examples of subsequent pollution.

Besides global warming—which could cost us the future of our children and our planet—other reasons compel us to kick the fossil fuel habit.

- Ocean acidification—harming ocean ecosystems
- Acid rain—killing forests and damaging property
- Smog and ground-level ozone—causing respiratory problems
- Mercury—harming humans and animals
- Death tolls

## GROWING ACIDITY of the WORLD'S OCEANS

*Global cost: Trillions of dollars annually—destruction of the ocean ecosystems and food chains. Risk: Are we willing to bet the planet's health and say goodbye to coral reefs?*

Fossil-fuel emissions from coal-burning power plants and gasoline powered automobiles make oceans and waterways more acidic. The gases dissolve in water to form acidic water and acid rain. Some scientists and policymakers interviewed by Juliet Eilperin of the *Washington Post* believe that acidification of the world's oceans could be the *most* pressing environmental threat facing earth. Noted expert Thomas E. Lovejoy rewrote a paperback edition of *Climate Change and Biodiversity* to highlight the threat of ocean acidification. "It's the single most profound environmental change I've learned about in my entire career," he said. Scientists warn that by the end of this century, the rising acidity of oceans could destroy the world's coral reefs and the creatures that underpin the sea's food chain. The problem is global and chronic. "What we do in the next decade will affect our oceans for millions of years," says Ken Caldeira, a chemical oceanographer at Stanford University. "$CO_2$ levels are going up extremely rapidly, and it's overwhelming our marine systems." Some ocean organisms probably lack the ability to adapt. Others can form only thin shells and weak skeletons.

This alone is reason enough to eliminate the use of fossil fuels immediately. Its consequences for oceans are certainly as serious as some predict for global warming. And the effects of fossil fuels on oceans is more easily studied, measured, and proven. But what is actually getting done?

"What we do in the next decade will affect our oceans for millions of years."

Congressional Representative Jay Inslee (D-Washington) declares acidic oceans have become "an absolute time bomb that's gone off both in the scientific community and, ultimately, in our public policy making." Inslee, after receiving a 2-hour briefing on ocean acidity in May 2006 with several other House members, said, "It's another example of when you put gigatons of carbon dioxide into the atmosphere you have these results none of us would have predicted."

John Pickrell of *National Geographic News*, reporting on two studies co-authored by Christopher Sabine, a geophysicist with the

National Oceanic and Atmospheric Administration, states that "Around half of all carbon dioxide produced by humans since the industrial revolution has dissolved into the world's oceans—with adverse effects for marine life." Greenhouse gases absorbed by oceans "are now changing ocean chemistry. The resulting change has slowed the growth of plankton, corals, and other invertebrates that serve as the most basic level of the ocean food chain. The impacts on marine life could be severe, scientists say." Sabine declares, "The oceans are performing a great service to humankind by removing this carbon dioxide from the atmosphere… The problem is that this service has potential consequences for the biology and ecosystem structure of the oceans." Recent articles by Sean Markey and Scott Norris estimate that ocean acidity has risen by 30 percent over the last 150 years, thus relieving somewhat the threat of $CO_2$-caused global warming, but greatly harming ocean ecosystems and food chains.

Hmmm, perhaps as some have suggested we can improve air quality by pumping more $CO_2$ into the oceans. Oops—this troubling suggestion could lead to another case of unintended consequences. Clean the air by fouling the oceans? I don't think marine life would like it.

> Ocean acidification is "the single most profound environmental change I've learned about in my entire career."

Although some well-qualified scientists question how rapidly or dramatically oceans will become more acidic, almost all scientists acknowledge that this phenomenon is easier to measure and model than global warming. Accurate measurements are possible, including measurements unique to specific locations. Stanford University marine biologists conclude that some coral reefs are "dissolving." Chemical oceanographer Ken Caldeira declares that "by the end of the century, no water will be as alkaline [salty] as where they [coral reefs] live now." If $CO_2$ emissions continue at their current levels, "It's say goodbye to coral reefs," said Caldeira. This is a clear, undeniable, measurable consequence of using fossil fuels.

## ACID RAIN

*Global cost: Trillions of dollars annually—destruction of countless plant and animal species, buildings, and art treasures. Risk: Are we willing to alter the natural and human landscapes of the planet and turn water into lemon juice?*

Sulfur dioxide ($SO_2$), nitrous oxide ($N_2O$), and other emissions from power plants—particularly coal-fired power plants—combine with water in the atmosphere to form acids that blow with the winds and fall to earth as rain, snow, sleet, hail, and fog. Acid precipitation, called acid rain, can be as acidic as lemon juice. Scientists measure acidity on a pH scale: 1 (highly acidic) to 7 (neutral) to 14 (highly alkaline). Rain is normally slightly acidic, with a pH of about 5.6, yet acid rain usually has a pH of 4 or 5, but can be lower. Acid rain causes massive and incalculably expensive damage to bridges, buildings, monuments, and forests, and it turns lakes and streams more acidic (lower pH), often killing all fish. Its effects are broad and diverse:

Acid rain can be as acidic as lemon juice.

- Acid-sensitive streams in New York's Catskill and Adirondack Mountains are becoming too acidic to support a diversity of life.

- Acid-sensitive lakes in New York, Canada, southern Norway, and Sweden cannot support important groups of insects, fish, frogs, toads, amphibians, and aquatic plants, which cannot survive and reproduce. As these species die or migrate, larger animals that rely on them also suffer. These animals include waterfowl and otters.

- Atlantic salmon populations will continue to decline in Nova Scotia.

- Reductions in fish diversity will persist in northwest Pennsylvania.

- There will be a continued decline in vigor of red spruce and sugar maple throughout the eastern United States and Canada.

## The Trail of Destruction

During the 1970s approximately 30 million tons of sulfur dioxide and 25 million tons of nitrous oxide were emitted into U.S. air each year. In the mid-1980s, Canada, Norway, and Sweden each declared acid rain their most serious environmental problem. By 1996 only about 20 million tons of $SO_2$ were emitted annually in the United States, but the emission of nitrogen oxides did not change. U.S. power plants still spew about 10 billion gallons of acid into the air every year. If distributed evenly over the entire area of the United States, then the 10 billion gallons amount to about 5 gallons per acre per year, but it doesn't work that way. The South and Midwest in the United States have the largest number of coal-fired power plants, but they escape the worst acid-rain damage, because winds carry this acid rain and dump it on the eastern coast of the United States and Canada. In some parts of the world the concentrations get very high. Want to know where? Just follow the trails of damage, disease, and death. In the United States, follow the trucks of contractors repairing damage, worth about $2 billion per year, to buildings and stone structures. In the Black Forest region of Germany, look for the dead trees. In Norway, follow the smell of dead fish. In Italy, follow the trail of marble statues adorned with bumps where noses and ears should be.

German forests are dying. In 1985, observers noted damage to over 85 percent of silver firs, about 60 percent of pines, and over 50 percent of spruces, oaks, and birches. The problem gets worse each year. Unofficial estimates put the cost of the damage at approximately $10 billion *per year*. Similar damage affects the forests of France, Sweden, Norway, the Czech Republic, Slovakia, Poland, Hungary, and Russia. The loss of these trees severely reduces the $CO_2$ absorbed by European forests, thus adding greatly to the $CO_2$ problem.

Forest damage is not confined to Europe. A study by scientists at the University of Vermont demonstrates that half of the spruce trees in the Camel's Hump area of Vermont's Green Mountains died between 1965 and 1981. Canada's $20-billion-per-year forest industry has also been damaged by acid rain.

Acid rain washes into rivers and streams, killing small aquatic organisms at the base of the food chain. As acidity increases, larger and larger fish die. The U.S. Office of Technology Assessment estimates that approximately 17,000 lakes and 112,000 miles of streams in the United States have been damaged by acidification. Canadian officials estimate that about 20 percent of the 50,000 lakes in eastern Canada have already acidified to threatening levels.

## The Height of Folly

To solve the problem of increasingly high concentrations of dangerous sulfur dioxide in cities, utility companies and coal-burning industries built ever-taller smokestacks to carry pollutants higher aloft and downwind to the less populated countryside. How generous. "Dilution is the solution to pollution" became the mantra. Engineers confronted a dilemma: How to provide more energy and reduce urban pollution levels while using polluting fuels? The following on smokestacks comes from Robert Morris' fine book *The Environmental Case for Nuclear Power* (Paragon House, 2000).

By 1981, 179 U.S. smokestacks stood over 490 feet in height. Twenty towered at least 980 feet. Industries in England, Germany, and other European countries also resorted to tall smokestacks. Sulfur dioxide levels dropped near the smokestacks. In industrialized Manchester, England, engineers reduced winter smoke levels by 90 percent and dropped sulfur dioxide levels by two-thirds, thereby doubling winter sunshine hours and cutting the death rate from bronchitis by half between 1956 and 1987.

Engineers had scarcely finished celebrating when data on downwind pollution began pouring in. The practice shared and distributed pollution, but it did not eliminate it. Sulfur dioxide produced by coal-burning power plants in Ohio, Indiana, Illinois, Pennsylvania, and West Virginia falls on eastern Canada. Acid rain from these states falls in the northeastern United States.

## SMOG and GROUND-LEVEL OZONE

*Global cost: Billions of dollars annually—human health problems and death, and crop losses. Risk: Are we willing to make gas masks and surgical masks every-day attire as we turn the very air we breathe into an atmospheric cesspool? Any cures yet for heart disease, lung cancer, emphysema, bronchitis, or asthma?*

The word *smog*—a combination of smoke and fog—was coined in London in 1905, but the problem has been around for much longer. Because of smoke and smog, London was once known as "The Smoke;" Edinburgh, as "Auld Reeky." Chinese cities are now earning similar nicknames. Consider Handan, as described by Joseph Kahn and Mark Landler in the *New York Times* in 2007.

> Residents of Handan live in a miasma of dust and smoke that environmental authorities acknowledge contains numerous carcinogens. After public protests, the company [Hangang Steel] agreed to pay an annual "pollution fee" to compensate some neighbors…. Tian Lanxiu claims "Hangang knocks 10 years off people's lives… Airborne concentrations of sulfur and benzopyrene, a byproduct of coking [and steel-making] [are] 100 times the levels measured in London."

Classic smog results mostly from burning coal. Smog is worst during hot summer weather. Photochemical smog, a very bad secondary pollutant, results from a reaction of sunlight and various toxic emissions, arising primarily from burning fossil fuels. Natural events, such as erupting volcanoes and long-burning forest fires, also cause smog.

Some smog is created from nitrous oxides and volatile organic

Smog burns your eyes and lungs, inflames breathing passages, decreases lung capacity, causes shortness of breath, induces wheezing and coughing, dries out the membranes of the nose and throat, causes pain when inhaling, and interferes with the body's ability to fight infection.

compounds (VOCs) released by motor vehicles and coal-burning power plants. VOC vapors can also be released from many other sources, including raw gasoline, alcohol, paints, solvents, pesticides, acetone, and benzene. Vapors also rise from, for example, formaldehyde (used in various building materials) and perchlorethlene (one of the world's worst carcinogens, used in dry cleaning clothes).

Wind-borne smog affects rural areas, but smog primarily affects cities, especially cities in geologic basins encircled by hills or mountains. Some of the great cities of the world—London, Los Angeles, Mexico City, Houston, Toronto, Athens, Beijing, and Hong Kong—have experienced dangerous levels of smog, even killer smog. In 1952, smog killed over 12,000 people in London. London has partially solved this problem by curtailing the burning of coal and by permitting the use of smokeless fuels only. However, smog caused by traffic pollution occurs today in London. In some Chinese cities people have trouble seeing across the street through dense smog and the tears in their burning eyes. Mexico City suffers terribly from smog and air pollution, because it is located in a geographic bowl of sorts. In a span of 30 years, the city's very clean air became among the worst—polluted in the world. Nitrous oxide concentrations in Mexico City are 2–3 times higher than recommended by international health standards.

Smog burns your eyes and lungs, inflames breathing passages, decreases lung capacity, causes shortness of breath, induces wheezing and coughing, dries out the membranes of the nose and throat, causes pain when inhaling, and interferes with the body's ability to fight infection. It causes long-term health problems, particularly for senior citizens, children, and those with heart and lung conditions such as emphysema, bronchitis, and asthma.

In 1952, smog killed over 12,000 people in London.

The natural ozone layer in the stratosphere shields the earth from harmful ultraviolet rays. Ground-level ozone, a powerful secondary pollutant and respiratory irritant, results from the reaction of ultraviolet light and some primary pollutants emitted by the burning of coal. Ground-level ozone triggers more than 6 million asthma attacks each year, destroys about 7 percent of U.S. agricultural production, and kills trees and plants. Ground-level ozone also interferes with the ability of plants to produce and store starch, thus weakening the plants, reducing their growth rates, and making them more susceptible to insects, disease, and other environmental stresses. Cutting present ozone concentrations in half would save $2 billion worth of crops per year in the United States.

## MERCURY

*Global cost: Trillions of dollars annually—brain damage in fetuses and infants, severe health threats to humans, mercury-laden fish unfit to eat. Risk: Are we willing to continue to foul air and water permanently while also threatening fetuses and newborns in the name of comfort and economic growth when we don't have to?*

Coal-burning power plants are the largest sources of acid rain and human-caused mercury emissions. When coal burns, mercury and other heavy metals vaporize into the environment. Wind, rivers, and oceans then transport the poisons around the globe, often traveling thousands of miles from where they are emitted. The mercury then ends up in the tissue of plants, animals, and humans. Mercury entering the atmosphere in Texas, Togo, or Turkmenistan could end up in the brain of a child in Nebraska, Nepal, or anywhere else in the world. Mercury is a particularly sinister pollutant because you cannot see or smell it, and it does not break down.

Acid rain may aggravate the mercury problem. High acidity in rivers, lakes, and streams appears to trigger releases of mercury from soils and sediments and the conversion of elemental mercury to a more dangerous compound of mercury—methyl mercury.

> One drop of mercury can contaminate a 25-acre lake and make its fish unfit to eat.

Unless control technologies are widely deployed and alternative energy sources introduced, mercury emissions will increase as the burning of fossil fuels increases to meet the rising energy and industrial demands of developing and developed nations. The United Nations Environmental Program (UNEP) in 2003 reported the conclusions of an international team of experts: Coal-fired power plants and waste incinerators now annually produce about 1500 tons (70 percent) worldwide of new, human-generated mercury emissions. Developing countries produce the majority of these emissions; over half (860 tons) comes from Asia. According to the U.S. Environmental Protection Agency in 2005, U.S. power plants annually spew 50 tons or more of mercury. However, just one drop of mercury can contaminate a 25-acre lake and make its fish unfit to eat.

Bacteria in streams and lakes convert elemental mercury, the metal in thermometers, to methyl mercury, a dangerous neurotoxin. It is ingested by small marine life, which are eaten by small fish, which are then eaten by progressively larger fish. Throughout the food chain the level of methyl mercury becomes more concentrated. Since mercury bonds strongly to protein, it is neither broken down nor easily eliminated. The concentration of methyl mercury in large fish can be over a million times higher than the water in which they lived. As birds, land animals, and humans eat the fish, the concentrations of mercury become more potent. Ingested methyl mercury travels freely throughout the body and can cross the blood-brain barrier.

Mercury can cause damage to brain, heart, lungs, kidney, thyroid glands, digestive system, liver, reproductive organs, nerves, muscles, and skin. Mercury also kills and causes insanity. For example, hatters in England, who used mercury to shape hats, often became insane. Their sad experience gives rise to the phrase "mad as a hatter" and to the character "The Mad Hatter" in Alice in Wonderland. In adults methyl mercury poisoning is associated with an increased risk of heart attacks. Effects on the brain can include irritability, tremors, harm to vision, memory loss, and impaired concentration. Some researchers suspect that mercury exposure plays a role in the development of Parkinson's disease, multiple sclerosis, Alzheimer's disease, and autism.

Methyl mercury's favorite targets are unborn babies and young people. It destroys their brains. This is why pregnant women should not eat fish. Methyl mercury is particularly dangerous to the nervous systems of fetuses and their still-forming brains. Affected children have slower reflexes, difficulty in learning, and shorter attention spans.

P. Bolger and B. Schwetz report in the *New England Journal of Medicine* that 7–8 percent of pregnant women have more methyl mercury in their bodies than doctors recommend. According to Jennifer Lee of the *New York Times*, the U.S. Environmental Protection Agency estimates that about 630,000 babies born each year in the United States could be at risk of brain damage from mercury poisoning. The U.S. Centers for Disease

> Methyl mercury's favorite targets are unborn babies and young people. It destroys their brains. This is why pregnant women should not eat fish.

Control and Prevention put the number of at-risk babies in the United States at about 300,000. Globally, the number could run into the millions. In northern Greenland, where people eat a lot of fish, 16 percent of the population has toxic levels of mercury in their blood. Could this be the reason the world has increasing needs for special education classes? Suspicious coincidence?

## What Can We Eat and How Much?

Predatory fish at the top of the food chain are generally more contaminated. These include oceanic and freshwater fish such as barracuda, burbot, eel, king mackerel, marlin, perch, pike, scabbard, shark, swordfish, tuna, and walleye. In southern and central Finland, an estimated 85 percent of pike weighing 2 pounds or more have methyl mercury concentrations that exceed international health limits. In Sweden, 60 percent of the approximately 100,000 lakes have pike with mercury levels that exceed international health standards.

The *U.S. Geological Survey Factsheet, #146–00* (October 2000) says that the steadily increasing number and geographic extent of state advisories against the consumption of fish because of mercury contamination has raised the awareness of the widespread nature of the mercury hazard. Fish consumption advisories for methyl mercury now account for more than three-quarters of all fish consumption advisories in the United States. Forty states have issued advisories for methyl mercury on selected water bodies, and 13 states have statewide advisories for some or all sport fish from rivers or lakes. Coastal areas along the Gulf of Mexico, Maine, and the Atlantic Ocean from Florida to North Carolina are under advisories for methyl mercury for certain fish.

**Risks to Wildlife**

Wildlife that eat fish risk contaminating themselves—they can't read the advisories.

In Wisconsin, loons lay fewer eggs when the eggs contain mercury in excess of concentrations that are toxic in laboratory studies. In part of the Everglades, the behavior of juvenile great egrets is affected if they ingest mercury in excess. Biochemical analyses show that mercury adversely affects diving ducks from the San Francisco Bay area, herons and egrets from Nevada's Carson River, and heron embryos from colonies along the Mississippi River. Studies with mallards, great egrets, and other aquatic birds also show harm following exposure to mercury.

Other contaminants also affect the toxicity of mercury. Methyl mercury can be more harmful to bird embryos when selenium, another potentially toxic element, is present in the diet.

**Mercury Worldwide**

The UNEP (2003) reports that the global threat from mercury to humans and wildlife has not diminished, despite reductions in mercury emissions in the economically advanced countries, because discharges from developing countries are rising rapidly. In some locales the problems are worsening as demand for energy rises.

## OTHER NASTY STUFF

The burning of coal annually dumps into the atmosphere at least 1000 tons each of beryllium, manganese, selenium, and nickel, and smaller amounts of lead, arsenic, cadmium, and asbestos. Does it surprise you that coal-burning power plants release much more radioactivity per day than nuclear power plants do? Yep, it's true. Consider this striking quotation from Robert Morris' fine book *The Environmental Case for Nuclear Power*, page 45: "Coal ashes are roughly 180 times more radioactive than

> Coal-burning power plants release much more radioactivity per day than nuclear power plants.

the level permissible for nuclear power plants." Morris further quotes Walter Marshall, chairman of the United Kingdom's Central Electricity Generating Board (CEGB):

> I have to inform you that yesterday the C.E.G.B. released about 300 kilograms (660 pounds) of radioactive uranium, together with all of its radioactive decay products, into the environment. Furthermore we released some 300 kilograms of uranium the day before that. We shall be releasing the same amount of uranium today, and we plan to do the same tomorrow. In fact, we do it every day of every year so long as we burn coal in our power stations. And we do not call that 'radioactive waste'. We call it coal ash.

## FOSSIL FUELS as MASS MURDERERS

Each year, approximately 50,000 Americans die from pollution caused by burning fossil fuels. Studies from Abt Associates—partially funded by the U.S. EPA and the Clean Air Task Force—report that about 24,000 of these deaths are attributable to pollution from fossil-fuel power plants. It seems clear to analysts that the other half died from auto emissions. Around the world, at least 2 million people per year die from fossil-fuel pollution, particularly coal. According to Robert Morris' book (pages 4–5),

> The combustion of coal and oil derivatives produces sulfur dioxide, carbon monoxide, mercury and other heavy metals, airborne ash, nitrogen oxide, poisonous radiation, and many cancer causing substances. When these dangerous air pollutants, even at relatively low levels, are breathed over a long period of time, they add to the death toll claimed by bronchitis, emphysema, lung cancer and heart disease.

Notice $CO_2$ does not appear on Morris' list of "serious" polluting substances.

Let's look carefully at the "death toll" by considering chronologically the following "crime reports" from around the world:

▶ *Earth Policy Institute* (B. Fischlowitz-Roberts, September 17, 2002)—"Air Pollution Fatalities Now Exceed Traffic Fatalities by 3 to 1." Reports from the World Health Organization (WHO) confirm that 3 million people in the world die each year from the effects of air pollution. This is 3 times the 1 million killed each year in automobile accidents. A study published by N. Künzli and colleagues in *The Lancet*, an esteemed British medical journal, concludes that air pollution in France, Austria, and Switzerland caused more than 40,000 deaths annually in those countries. About half of these deaths are traceable to air pollution emitted from vehicles. In the United States, annual traffic fatalities total over 40,000, while air pollution claims at least 50,000 lives. *U.S. air pollution deaths equal deaths from breast cancer and prostate cancer combined.*

The fact that air-pollution fatalities substantially exceed traffic fatalities worldwide demands a broad redefinition of "auto safety" to include decreasing air pollution.

The fact that air-pollution fatalities substantially exceed traffic fatalities worldwide demands a broad redefinition of "auto safety" to include decreasing air pollution. While only some motorists contribute to traffic fatalities, all motorists contribute to air pollution fatalities.

▷ *World Health Organization* (Ezzati et al., 2004; Ostro, 2004; Krzyzanowski, 2007)—Air pollution in urban areas contributes to the death of 1.5 million people in Asia every year. Pollution-caused diseases are among the most debilitating and painful—aggravated asthma, bronchitis, emphysema, lung and heart disease, and respiratory allergies.

These problems will accelerate as urban populations grow rapidly in coming decades.

▷ U.S. Environmental Protection Agency, (study prepared by Abt Associates and reported by J.R. Pegg in 2004)—The EPA's air quality consultants, using standard EPA methodology, found that:

• Pollution from power plants cuts short the lives of nearly 24,000 Americans nationwide every year. Those 24,000 Americans die an average of 14 years early. Lung cancer causes 2800 of those deaths annually.

• Power-plant pollution causes 38,200 non-fatal heart attacks per year.

▷ *BBC News* (July 2004)—"Polluted air from America could be damaging the health of people in Britain, experts fear."

Here is a clear demonstration of the global scope of pollution and the need for a global solution. One attempt at the solution, the Kyoto Protocol, is laughably feeble and is, at best, only a Band-Aid on a gaping wound.

▷ *BBC News* (February 2005)—"Air pollution [in Europe] is responsible for 310,000 premature deaths each year," according to a study by the European Commission. The study further states that pollution-related illnesses cost the European economy more than 80 billion Euros ($100 billion U.S.) per year.

▶ *The Metro West Daily News* [Framington, Massachusetts], (J. Brodkin, February 23, 2005)—"Diesel pollution is responsible for more deaths than drunk drivers and homicides, according to a new study that estimates how many premature deaths, asthma attacks and heart attacks are caused by diesel pollution in every U.S. County."

▶ *BBC News* (April 2005)—"The European Union could save 161 billion Euros [$200 billion U.S.] a year by reducing deaths by air pollution." According to the World Health Organization (WHO), the major contributors to air pollution are fossil-fuel-burning engines and the use of fossil fuels to heat and cool homes.

The "sticker price" we pay for fossil fuels is minute compared to the "real" cost in death, disease, and destruction. Consumers "subsidize" fossil-fuel industries by paying out of their pocket the costs for healthcare and military security. These costs are necessary in part because of the fossil-fuel industries' products and processes. We subsidize fossil fuels with a portion of our lives: an average of 6–12 months for each of us when figuring the 3 million premature deaths each year from airborne pollution. We also subsidize fossil fuels by the costs we pay—and our children and grandchildren will pay—for health, security, and environmental devastation. You never thought of the defense budget and your cancer treatments as energy subsidies, did you?

> The "sticker price" we pay for fossil fuels is minute compared to the "real" cost in death, disease, and destruction.

## REGULATION IS NO REMEDY

Regulatory groups in the United States try to muzzle pollution-belching, coal-fired, electricity-producing plants. Yet regulation and enforcement are often lax. The utilities clearly understand that hefty campaign contributions can produce legislation that lets them literally get away with killing people. Further, as long as regulatory fines cost less than the profits reaped, then firms will pay the fines as a cost of business. Utilities just want to keep costs down, but low costs often mean high death rates.

In the last 30 years, the U.S. Congress, individual U.S. states, and some European countries enacted clean air standards to strengthen the regulation of air pollution. These standards set numerical limits on the concentration of some air pollutants and create reporting mechanisms. This is good, but not nearly good enough—again, like a finger in a dike. Pollution marches on, but at a slower pace.

Yes, scrubbers and other devices in coal-fired plants have lessened some pollutants, and some new plants greatly reduced the volume and number of pollutants released into the atmosphere. Yet many utilities chose to expand some existing plants rather than build new ones, because the old plants were not required to conform to the same strict regulations imposed on new ones. Why don't regulations apply to all plants, old and new? Forget complex answers to this question—I say follow the money.

## BOTTOM LINE

Savings in healthcare costs alone can fully support the transition to a clean, renewable, eternal energy future. Now add the cost of fish killed by ocean acidification, forests destroyed by acid rain, and other fossil-fuel-caused calamities, and I guarantee that the transition will cost the U.S. public nothing. In fact, I will bet on it—any takers? The hard money cost/benefit ratio makes the transition even more painless. See Energy Independence (Chapter Sixteen) for details.

- There is absolutely no way to combat the many fossil fuel disasters, except to abandon the cause, fossil fuels themselves. The situation is desperate since many of the problems can become chronic or irreversible. We have the technology and the money to correct these problems, but I honestly don't know if we have the time.

- Forget quick fixes. Forget fix-it-up schemes. We need to use the fundamental solutions already available. We know what they are, and we can easily afford them. Awareness, new ideas, inventions, advocacy, and political will are essential. Time is running out.

- Lots of numbers here, but the message is clear and unavoidable. We must get going toward a comprehensive solution.

**Note:** Much of the data and statistics on energy matters are in conflict. The data sometimes is *vastly* different, often from the same source. Presented with conflicting data, one must dig deeper to determine which numbers make the most sense, a very time-consuming exercise. Some differences can be attributed to data from different years and assumptions made in constructing the data. Still, some analysts seem to have pulled numbers out of thin air.

# Chapter 3

## POPULATION

Population growth is the single greatest reason for the accelerating depletion of raw materials, is the single greatest cause of increased energy use, and is the most formidable challenge in the world's quest to clean up the environment. The present global population growth rate makes it practically impossible to manage world health, food supplies, pollution, and poverty. Population growth and subsequent pressure on resources will certainly trigger violence and "resource wars" over scarce water, oil, arable land, timber, and the like. The magnitude of the problem should terrify us.

You think China has a lot of people? Consider this: If the world continues reproducing as predicted, then global population will add the equivalent of 2.5 more Chinas in the next 30 years. The pressure on all resources will be staggering. Many natural resources will simply

The present global population growth rate makes it practically impossible to manage world health, food supplies, pollution, and poverty.

run out. Other resources, such as food, will continue on a downward spiral as land and sea resources are depleted or destroyed.

World population in March 2008 was approximately 6.65 billion. How fast will the population grow? As one of many sources declared, such numbers are "not uncontroversial." I'm saying about 9.5 billion in 30 years. The global growth rate from 1975–2006 was about 1.5 percent annually, in spite of the many people around the globe who died of AIDS, other diseases, and wars. Recent U.S. population growth has been about 1.25 percent annually.

Let's say that population growth slows, as many predict. The current global population of 6.65 billion at a growth rate of 1.25 percent over 30 years yields a global population of 9.65 billion in 2038. Now let's take the U.N.'s estimate that the current population growth rate is 1.15 percent annually. Over 30 years that still yields a worldwide population of 9.4 billion. So let's just say 9.5 billion.

Other projections predict faster or slower *rates* of growth, but they still anticipate continued substantial growth in population *totals*. A slower rate of growth merely means it will take longer to get to 10 billion people. Still, almost all credible projections predict 9–10 billion people by 2050 at the latest. When all is said and done, does it really matter if the world hits the 10 billion mark in 2055 rather than 2040 or 2045? By *any* measure population growth is out of control.

An increasing population demands increasing amounts of energy and fossil fuels. That means the resources will deplete more quickly and the pollution will pile up faster.

## POPULATION DEMOGRAPHICS

According to the Population Institute and other sources, more than 80 percent of the world's population live in developing countries, and 40 percent of those are children and teenagers either coming into or in their prime reproducing years. Many of these people live in poverty.

Nearly 4 out of 5 people in the world are considered poor, existing on less then $10 per day. Half of India's population of almost 1 billion lives on less than $1 per day. Their current poverty means they do not yet consume many resources per person. Yet that is exactly what they will do as economic development and industrialization occur in those countries. In contrast, population growth in the world's richest countries, excluding the United States, is near zero.

Many countries are trying to curb their population growth. China has a one-child-per-couple policy, with tough penalties for violators. However, population growth in China is still substantial. According to Alan Weisman in *The World Without Us*, as reported by Jerry Adler in *Newsweek*, if each couple in the world were limited to 1 child, then the world population would stabilize at 1.6 billion by the end of the century, where it was in 1900. India's population will likely grow between 1.45 and 1.8 percent per year from an already large, young population base. India and many other countries are already running out of natural resources due to population pressures. Water is a huge problem, crop harvests are down, and wood fuel is disappearing. Forests are cut to make way for food, and many countries that had been exporters of food are now importers.

Not only is the population growing in China, India, and elsewhere, but so are people's appetites for meat, private cars, and a better way of life. These changing appetites, tastes, and preferences in turn cause people to use more water, oil, and coal, to cut down more trees, and to dump more waste. Who can blame them—they work hard, and in some ways they are trying to emulate the "American dream" of years past.

## Historical Growth in World's Population

The world's population is growing exponentially.

1800—The world's population reached 1 billion

1930—130 years later: 2 billion

1960—30 years later: 3 billion

1975—15 years later: 4 billion

1987—12 years later: 5 billion

1998—11 years later: 6 billion

2007—9 years later:  approximately 6.6 billion

## Population Growth Varies

Population growth rates vary for several reasons (see Figure 3.1).

- The status and education of women in a given society: Population growth rates are low in advanced countries, while high in countries where women have low status and poor education.

- Lack of financial security for old age: Couples have more children to support them, and children are an economic asset in some societies.

- Availability of family planning services.

- Access to family planning, and contraception.

Low population growth rates come with education and advanced societies. Mass education programs would be hugely helpful. Yet widespread education and social advancement may take centuries, unless countries have adequate clean energy to grow and prosper. That situation is very unlikely with the current population explosion and even less likely since prosperous countries have a serious energy problem themselves. Modern wealthy countries could start by providing at least adequate electrical energy to poor countries. This would be a good investment that wealthy countries can afford—helping others to help themselves. A higher standard of living and education produces better population control.

## Figure 3.1. The 15 Most Populous Countries in 2008 and Their Annual Population Growth Rates

| | Country | Population 2008 | Annual Population Growth Rate* (Percent) | Estimated Population 2038 |
|---|---|---|---|---|
| 1 | China | 1,300,000,000 | 0.69 | 1,600,000,000 |
| 2 | India | 1,100,000,000 | 1.45 | 1,700,000,000 |
| 3 | U.S. | 300,000,000 | 0.90 | 390,000,000 |
| 4 | Indonesia | 242,000,000 | 1.57 | 386,000,000 |
| 5 | Brazil | 186,000,000 | 1.06 | 255,000,000 |
| 6 | Pakistan | 162,000,000 | 2.17 | 308,000,000 |
| 7 | Bangladesh | 144,000,000 | 1.85 | 250,000,000 |
| 8 | Russia | 143,000,000 | -0.37 | 120,000,000 |
| 9 | Nigeria | 129,000,000 | 2.37 | 260,000,000 |
| 10 | Japan | 127,000,000 | 0.05 | 129,000,000 |
| 11 | Mexico | 106,000,000 | 1.17 | 150,000,000 |
| 12 | Philippines | 88,000,000 | 1.84 | 152,000,000 |
| 13 | Vietnam | 84,000,000 | 1.04 | 115,000,000 |
| 14 | Germany | 82,000,000 | 0.00 | 82,000,000 |
| 15 | Egypt | 78,000,000 | 1.78 | 132,000,000 |
| | **Rest of World** | 2,379,000,000 | 1.27 | 3,471,000,000 |
| | **World** | 6,650,000,000 | 1.25 | 9,500,000,000 |

* Growth rates vary greatly. Sources don't agree in detail, but all are comparable.

## HOW POPULATION GROWTH WILL AFFECT WORLDWIDE ENERGY USE

As population grows, both depletion and pollution will accelerate proportionately, especially if individuals and industries continue to burn fossil fuels, particularly coal. In fact, I believe if there is a global environmental tipping point so many have warned about, then it could happen as a result of the increased use of fossil fuels arising from predicted population growth. Devastation discussed elsewhere in this book would accelerate, as many toxic pollutants cumulate. The world might run out of fossil fuels in the midst of other disasters and thus lose the capacity to fix the problem. The math is simple. The consequences are dire. The world needs an immediate, crash program to migrate quickly to renewable energy sources.

## BOTTOM LINE

- You've heard the expression, "an ounce of prevention is worth a pound of cure." The world must find a way to curtail population growth. If we won't do it voluntarily, then Mother Nature will step in and do it her way—by plague, famine, war, and death. This perfect, apocalyptic storm is gathering strength. Population growth is a towering feature of this impending storm.

- The world must not give up on solving the population problem even though it seems hopeless, at least in the near term.

- Conservation, while it always makes sense, does little to solve the world's energy problems. As the world conserves, there are continually more users canceling out some or all of the positive effects of conservation.

# Chapter 4

## GLOBAL WARMING

### THE GREAT DISTRACTION

Global warming has unfortunately become *the* environmental and energy battleground. This focus is unfortunate because global warming is a distraction from the real peril. Also, scientists and experts will likely disprove some of the main reasons offered to explain global warming. If that happens, then people will relax and not recognize the bigger, incredibly serious, and complex problem that underlies global warming—the burning of fossil fuels. In the meantime attention to global warming simply deflects the public's attention from the perilous problems of fossil fuels.

The discussion or controversy over global warming has become an unnecessary part of the national and global energy problem, yet debates over global warming are confusing and unresolvable. The real problem is very simple: *the burning of fossil fuels at ever-increasing rates.*

Even though scientists do not agree on what causes global warming, it would be insane to bet against those who are convinced that all or most global warming is caused by human activities. If these scientists and pundits are right, then the consequences of business as usual will be catastrophic. Why tempt the devil? At the same time, those who contend that global warming arises in large part from natural causes are also very convincing. For now let's concede the argument and assume all global warming is caused by human activities. I can easily make this concession because human-caused global warming becomes a non-issue once the world tackles the real, underlying problem of the burning of fossil fuels. The only *certain* solution to the problem of human-caused global warming is to stop burning fossil fuels. It's that simple.

## MR. GORE, GLOBAL WARMING IS a SYMPTOM, NOT the PROBLEM

Mr. Gore, congratulations on your Nobel Peace Prize and other awards. Your commitment to the creation and presentation of *An Inconvenient Truth* is inspiring. The awards and acclaim provide you an unrivalled platform that may attract attention to the issues and spark action—at least that is my hope. You have become the de facto spokesperson for environmental security and energy reform. Despite my sincere admiration for your effort, I think your fundamental message is incomplete and one-sided, your solutions inadequate, and your position a distraction from the fundamental problem. It is time to tell it like it is.

> The only certain solution to the problem of human-caused global warming is to stop burning fossil fuels. It's that simple.

You draw attention to global warming, but global warming is merely one dramatic symptom—one symptom among many—of the problem of burning fossil fuels. Reducing a fever is not the same as remedying an illness or curing a disease. The real problem is fossil fuels. Their continued use means we will be running out of these fuels and accumulating too much pollution and environmental devastation, including global warming. The burning of fossil fuels is the world's colossal problem requiring a colossal, civilizational, global solution. Let me say it again: The only real, viable, certain way to end human-caused global warming is to stop using fossil fuels.

I worry that your declarations and acclaim focus the public's attention too much on the *apparent* problem of global warming and distract it from the *real* peril of fossil fuels. You imply that human activity causes 100 percent of global warming, and you further imply that all thoughtful scientists agree with your assessment. You offer an impressive score of 928 to 0— that is, 928 peer-reviewed scientific articles and papers agree with you that global warming is spinning out of control because human activities are belching massive amounts of carbon dioxide into the atmosphere. Zero disagree with you, you say.

These claims as grossly misleading. There is no doubt climate is changing. There is little doubt that humans have something to do with it, but many highly respected scientists around the world do not agree that human activities are the *major cause* of global warming or that atmospheric $CO_2$ will cause global warming to spin out of control. Also, there is no doubt that climate has changed from the beginning of time. Thus, considered opinions vary over whether the current experience of global warming is caused by natural cyclical changes or by the human-caused buildup of $CO_2$ and

> Just quit burning fossil fuels. It matters little why a train charges toward us or how fast it approaches. We must simply get off the tracks.

other greenhouse gases in the atmosphere. While there is a growing consensus that the earth is currently in a warming cycle, some very credible scientists insist that global warming is not universal, arguing that some places are warming while others are cooling. You might be surprised at how thoughtful and analytically sound some opposing points of view are.

Whether you or I believe human activity is primarily responsible for global warming, or is just contributing to it, is not relevant. Why not eliminate all doubt? Why place a risky bet we don't have to make? Why bet our children's futures? The burning of fossil fuels certainly depletes reserves, threatens economic and social calamity, belches poisons, and kills. That's insane. Forget global warming. Consider the *certain* threats. If we stop using fossil fuels, then we save ourselves from several looming disasters, and a better environment is an added benefit.

Ending the burning of fossil fuels ends the debate on global warming as well as other prominent problems: depletion, acidic oceans, acid rain, smog and ozone, and mercury poisoning.

I have no problem with media outlets or anybody else turning human-generated $CO_2$ into an environmental boogeyman responsible for severe floods, rising sea levels, drought, famine, disease, tornados, hurricanes, and deadly fires. Heck, blame dandruff, crabgrass, and junkmail on carbon dioxide, too, but do it only if such simplified blame causes us to end the burning of fossil fuels, particularly coal. The burning of fossil fuels is our great problem—a profound threat to our environment, our economies, and, indirectly, to our political system. All else is only a distraction.

Here's why: Only by ending the use of fossil fuels can we hope to decrease or end the creation of most greenhouse gases that allegedly foster global warming. Why get drawn up in debates over which gases, how much, who burns them, how much they pollute, and who should pay? Just quit burning fossil fuels. It matters little *why* a train charges toward us or *how fast* it approaches. We must simply get off the tracks. Climate change is simply one more reason—not *the* reason—to stop burning fossil fuels.

Ending the burning of fossil fuels ends the debate on global warming.

## GREENHOUSE GASES

Several gases in the atmosphere trap the sun's reflected heat, thereby creating the "greenhouse effect." In moderation, such heat trapping modulates the temperature of the earth to maintain an average global temperature of about 60 degrees Fahrenheit (15 degrees Celsius).

However, if the concentration of these gases gets too high, then they may trap too much heat and cause the earth to warm up. This is the theory. Yet I have not seen anywhere a meaningful graph showing how concentrations of these gases actually affect the earth's temperature.

The major greenhouse gases are water vapor, which causes 35–70 percent of the greenhouse effect, carbon dioxide (9–26 percent), methane (4–9 percent), and ozone (3–7 percent). The large variations allow for a lot of speculation, modeling error, and downright confusion. Try juggling these numbers, and you'll find an infinite number of combinations that defy, it seems, any attempt at accurate modeling.

Some sources report that $CO_2$ accounts for 21 percent of global warming, and other gases contribute 15 percent. Differing sources suggest that "other gases" could contribute as much as 38 percent. These discrepancies illustrate the limits of current scientific knowledge and prevailing climate models. Different sources give different contributing percentages and some give a range. They are all partially correct, depending on the assumptions. The result is a lot of uncertainty and a nightmare when trying to collect accurate data.

Greenhouse gases are not all equal. Some are more powerful greenhouse gases than others. They are rated on a scale of $CO_2$ equivalents. For example, nitrous oxide is about 300 times more efficient at trapping heat than carbon dioxide.

**Carbon Dioxide**

Within modest limits, carbon dioxide is not a pollutant. It occurs naturally in the air we breathe and helps keep the earth's temperature suitable for life. Carbon dioxide through the process of photosynthesis is a fundamental building block for all plant life. In excess, $CO_2$, like almost anything, can cause a problem.

Carbon dioxide is a long-lived gas. Scientists expect that it lasts in the atmosphere for about 100 years. Some think carbon dioxide is the foremost cause of global warming, yet others think its overall contribution to global warming is small compared to increases in solar output (see G. E. Marsh, "A Global Warming Primer"). Experiments show that the response of temperature to carbon dioxide is logarithmic, not linear, meaning that the effect of $CO_2$ on warming drops off as the concentration of carbon dioxide increases (see Patrick J. Michaels, *Meltdown* (Cato Institute)).

Carbon dioxide is also one of the primary gases produced by the burning of fossil fuels. How much $CO_2$ does the burning of fossil fuels produce? I repeat the answers because they are staggering.

- 1 pound of coal produces about 3.4 pounds of carbon dioxide, plus large amounts of mercury, sulfur dioxide, and nitrous oxide.
- 1 pound (gallon) of gasoline produces 3.1 (17) pounds of carbon dioxide, plus sulfur dioxide and nitrous oxide.
- 1 pound of natural gas produces 2.75 pounds of carbon dioxide.

The United States annually emits 7 billion tons of $CO_2$ (2007), yet by 2040 this total is expected to rise to 9 billion tons of $CO_2$ per year. The world emits 31 billion tons of $CO_2$ per year now, a total projected to rise to 55 billion tons of $CO_2$ annually by 2040. For perspective, 31 billion tons of pure $CO_2$ could blanket the entire state of New York to a depth of 500 feet. Remember, these annual emissions are over and above the natural $CO_2$ cycle.

Since the beginning of the Industrial Revolution in 1750, atmospheric concentrations of $CO_2$ increased 36 percent, from 280 parts per million (ppm) to 380 ppm today. Experts predict the concentration will further increase to over 500 ppm before the end of this century, at which time some predict run-away global warming will destroy life on this planet. I don't believe it. While I don't believe it, only a fool would bet against it.

## Natural Gas (Methane)

Methane ($CH_4$), the principle component of natural gas, is about 22 times more efficient at trapping heat than carbon dioxide, thus methane's "equivalence" is 22 times an equal volume of $CO_2$. However, methane does not stick around as long—only 10–12 years versus 100 years. Scientists writing in *Science, Nature,* and other journals in June 2006 report that permafrost—the permanently frozen ground in the northernmost latitudes—may contain vast amounts of carbon in the form of methane and carbon dioxide. As the earth warms and permafrost thaws, these greenhouse gases are released. About 500 billion tons of greenhouse gases could be released from melting permafrost. For comparison, the earth's atmosphere currently holds about 700 billion tons of greenhouse gases, according to Keay Davidson. The permafrost "reservoir" contains methane equivalent to more than 100 times the amount of carbon released annually by the burning of fossil fuels. The timing of this methane release is very speculative, as are many such assertions. When would it happen? How quickly?

In the United States the largest methane emissions come from the decomposition of waste in landfills (23 percent), ruminant digestion and manure management associated with domestic animals (28 percent), natural gas and oil systems (26 percent), and coal mining (11 percent). Cattle in the United States release from both ends about 160 million tons of carbon dioxide *equivalence* per year—now that is a lot of cow gas. While it is difficult to believe, this is equivalent to the amount of $CO_2$ produced by more than 20 million cars per year, but you don't hear much about this "natural" pollution. Another oddity of the global warming saga.

Atmospheric methane has increased from 0.7 ppm before 1750 and the onset of the Industrial Revolution to approximately 1.8 ppm today, representing a 150 percent increase, but the concentration has remained fairly constant in recent years. Does this fit into the modeling? Some estimate that atmospheric methane could account for up to 20 percent of human-induced global warming. Nobody really knows.

## Other Greenhouse Gases

Nitrous oxide ($N_2O$) is 300 times more efficient at trapping heat than $CO_2$, but luckily and mysteriously its concentration since 1750 has increased only 16 percent—from 0.27 ppm to approximately 0.31

ppm. Experts estimate this gas could account for about 6 percent of global warming. Again, nobody knows.

Other gases, such as chlorofluorocarbon (CFC) refrigerants and surface ozone, account for approximately 13 percent of human-caused global warming. One cubic foot of CFCs is equivalent to 1300 cubic feet of $CO_2$. How much is out there? What is its contribution to global warming? Once again, nobody knows.

## Water Vapor: The Mother of All Greenhouse Gases

Nobody debates the fact that water vapor is the most important of the greenhouse gases. *Some* estimate that it contributes up to 64 percent of the greenhouse gases that allegedly contribute to global warming. Its contribution to global warming seems to vary a great deal —from 4–10 times the contribution from $CO_2$—which is why $CO_2$ is actually a minor greenhouse gas. Concentrations of water vapor vary regionally, and human behavior does not directly affect the amount of water vapor in the air. Let me say this another way: *Humans cannot affect water vapor in the atmosphere, the biggest source of greenhouse gases.*

Some climate models predict an increase in the earth's temperature will cause increased evaporation, thus more water vapor in the atmosphere. This in turn leads to a further increase in the temperature. Should we expect a runaway condition? Why doesn't it happen? I've not found anybody who knows for sure. There are obviously compensating mechanisms.

## GLOBAL WARMING and the MEDIA

Take a look at these screaming headlines, titles, and foreboding quotations.

▸ *Time* magazine, cover stories, April 3, 2006

"Be worried, be very worried. By any measure, earth is at the tipping point." "The climate is crashing and global warming is to blame."

"Never mind what you have heard of global warming as a slow-motion emergency that would take decades to play out. Suddenly and unexpectedly, the crisis is upon us."

▶ *National Geographic*, cover stories, September, 2004

"Global Warning—Bulletins from a Warmer World"
"Signs from Earth Heating Up—Melting Down"

▶ *The Economist*, September 7, 2006

"The Heat is On: A Special Report on Climate Change"

▶ *An Inconvenient Truth*, the film and book by Al Gore

▶ The International Climate Change Taskforce, co-convened by the Center for American Progress, reported extreme predictions about climate change. Amy Ridenour of the watchdog group National Policy Analysis declared that the task force "cherry picked a compilation of seemingly every doomsday scenario advanced by global warming alarmists over the past decade."

There seem to be thousands of such articles, studies, and reports. Amid the furor, however, prominent questions and dissenting voices arise. *National Geographic* (September 2004) declares "There's no question that the Earth is getting hotter—and fast. The real questions are: How much of the warming is our fault, and are we willing to slow the meltdown by curbing our insatiable appetite for fossil fuels?" *The Economist* (September 2006) reports that arguments about climate change are

fueled by ignorance, because nobody knows for sure what is happening to the climate. At a macro level, modeling what is

**Arguments about climate change are fueled by ignorance.**

the world's most complex mechanism, and projecting 100 years ahead, is tricky. At a micro level, individual pieces of data contradict each other. One shrinking glacier can be countered by another that is growing; one area of diminishing precipitation can be answered by another's where it is rising.

An editorial in the *Wall Street Journal* (February 5, 2007), commenting on a recent report from the United Nations Intergovernmental Panel on Climate Change (IPCC), warns "Beware of claims that the science of global warming is settled." The *Journal* also reports that the IPCC appeared to be "backpedaling" on some key issues.

Hmmm, maybe there is actually a debate here with competing views. Aside from the limits of science, what are the issues?

## WHAT ARE SCIENTISTS and AL GORE SAYING ABOUT GLOBAL WARMING?

Let's consider the opinions of several widely respected and widely quoted scientists on global climate: James Hansen, Gerald Marsh, Patrick J. Michaels—and Al Gore. Dr. Hansen heads the Institute for Space Studies at the U.S. National Aeronautics and Space Administration. He is also adjunct professor in Earth and Environmental Sciences at Columbia University. Mr. Marsh, a retired physicist from the Argonne National Laboratory, is a fellow of the American Physical Society, a former consultant to the U.S. Department of Defense under the Reagan, Bush, and Clinton administrations, and a member of the National Center for Public Policy Research. Dr. Michaels is research professor of Environmental Sciences at the University of Virginia and senior fellow in Environmental Studies at the Cato Institute. In 2003 he won an award for public service writing and "Paper of the Year" honors for a scholarly article on climate science. Mr. Al Gore won the Nobel Peace Prize for raising the world's awareness of global warming issues.

## Dr. James Hansen

James Hansen scared all of us and set the scientific community on its collective heels with his provocative congressional testimony on climate change on June 23, 1988. In 2001 he stated,

> Future global warming can be predicted much more accurately than is generally realized.... [W]e predict warming in the next 50 years of 0.75 degree Celsius ± 0.25 degree Celsius (approximately 1.5 degrees ± 0.5 degrees Fahrenheit), a warming rate of 0.15 degrees Celsius ± 0.5 (0.25 degrees ± 0.1 degrees Fahrenheit) per decade.

He further stated that much of the warming of the next 50 years will result from emissions *already* in the atmosphere. This doesn't sound too bad. Modest warming from damage already inflicted.

Yet Hansen more recently declared his opinion that if the world continues with business as usual, then the earth will reach an environmental "tipping point" by about 2016, at which point global warming will become an unstoppable, runaway phenomenon—8 years to go.

Wait a minute. Which is it? Mild warming or runaway climate change?

## Gerald Marsh

Mr. Gerald Marsh wrote an informational paper entitled "A Global Warming Primer." The paper refers to $CO_2$ as an important *minor* greenhouse gas, not as the major cause of global warming. According to Marsh, since 1000 AD worldwide temperatures have varied over a range of about 1.5 degrees Celsius (2.7 degrees Fahrenheit). Over most of the last 10,000 years, temperatures varied within a range of 2 degrees Celsius (3.6 degrees Fahrenheit). Marsh considers this 2-degree range the "natural variation" over this time. Relative to the average temperature over the last 10,000 years, global temperatures between 1860 and 1980 varied as follows:

- ‣ 1860–1920—Global temperature was about 0.3 degrees Celsius (0.5 degrees Fahrenheit) cooler than the average.

▶ 1920–1940—Global temperature rose by about 0.35 degrees Celsius (0.6 degrees Fahrenheit) to slightly above the average. ($CO_2$ concentration was close to historical norms.)

▶ 1940–1975—Global temperature shows a gradual cooling of about 1 degree Celsius (1.8 degrees Fahrenheit), although $CO_2$ concentrations were rising during the period.

▶ 1975–1990—Global temperature rose above the average by about 0.3 degrees Celsius (0.5 degrees Fahrenheit.)

The total range of temperature variation since 1860 is within about 0.6 degrees Celsius (1.1 degrees Fahrenheit.) This is more than 3 times smaller than the natural range of 2 degrees Celsius (3.6 degrees Fahrenheit.)

Hold on. If over the past 10,000 years temperatures varied within a range of 2 degrees Celsius (3.6 degrees Fahrenheit), then did $CO_2$ levels vary accordingly as shown on Al Gore's graphs? No. Since the concentration of $CO_2$ over most of this long period is believed to be relatively constant, the temperature variations must be due to other causes—not $CO_2$ concentration. Based on his analysis, Marsh concludes it is not possible for human activity to produce a runaway greenhouse effect. Rather, variations in the intensity of solar radiation may explain much of the temperature change, and uncertainties in climate modeling explain much of the analytical confusion. Further, Marsh quotes Ahilleas Maurellis of the Space Research Organization of Netherlands. The quotation appeared in *Physics World* (February 2001):

> Ultimately it is too simplistic to blame global warming on a particular gas or process... Perhaps the real villain is not carbon dioxide or even water vapour, but simply a mixture of inertia, hysteria and misinformation. Until we

understand the full picture, perhaps the best reaction to global warming is for everybody to just keep their cool.

## Dr. Patrick J. Michaels

Patrick J. Michaels wrote the excellent book *Meltdown*, which begins with a succinct discussion of climate-change science and then unrolls a litany of climate-related falsehoods, exaggerations, and misstatements. He cites a multitude of errors and exaggerations in scientific papers, news reports, and television sound bites—from the "National Assessment of Global Warming," a Clinton-era document that used computer models its authors knew did not work, to the infamous *New York Times* stories, eventually retracted, about the melting of the North Pole (see Michaels, pages 42–46).

Michaels argues that the way scientists conduct research today—amid competition among projects for monopoly funding by the federal government—"creates a culture of exaggeration and a political community of saviors that takes credit for saving us from certain doom, whether it is the doom of omission or commission." Michaels declares that the Kyoto Protocol on climate change, which we've heard such ballyhoo about, would reduce surface temperatures in 50 years by only 0.07 degrees Celsius (0.13 degrees Fahrenheit). Yet even this won't happen since not one—*not one*—of the signatories is expected to meet the agreement's modest emissions reductions. This is a Band-Aid fix at best, even if all nations complied. It is a symbolic political victory of sorts—*see, we did something*—even if it is almost environmentally meaningless. And the United States, of course, did not sign. Further, the Kyoto Protocol's proposal to *reduce* emissions scares me. We must *eliminate* emissions, not reduce them. Reductions offer no solution. They only buy us a little time.

If politics is often reducible to symbols and theatre, to appearances and perception, to form over substance, then it shouldn't surprise us that some critics, including Michaels, sharply rebuke the United Nation's Intergovernmental Panel on Climate Change as a producer of slick, "political" volumes on climate assessments and predictions. These reports, considered by many to be state-of-the-art studies, are the product of hundreds of scientists and reviewers, who could be described as a collection of bureaucrats. Further, the studies often ignore measurements or data that don't agree with their conclusions. Does the IPCC trade on its prestigious name and level of public trust to contribute to the hysteria on global warming? For example, one IPCC study concludes that global temperatures could rise by 5.8 degrees Celsius (10.4 degrees Fahrenheit) in 100 years. Many climatologists don't believe this is remotely possible. Please refer to the books and papers I cite, and you be the judge.

## Other Scientists Around the World

Are Marsh, Michaels, and others who dare to question the present global-warming hysteria just extremists and climate heretics? What do other scientists around the world think? Let's look at a 2003 survey on global warming (copyrighted and published in 2007) conducted by two German environmental scientists, Dennis Bray and Hans von Storch. Bray is a research scientist at the GKSS Institute of Coastal Research in Geesthacht, Germany. Von Storch is a climatology professor at the University of Hamburg and director of the Institute of Coastal Research. Participants in the survey included 520 climate scientists from 27 different countries. The following graphs (see Figure 4.1) are responses to 4 of the most pertinent questions:

1. Are humans causing climate change?
2. Can scientific models predict future climate?
3. Does the IPCC reflect scientific consensus?
4. Can we predict climate variability on time scales of 10 years?

It seems there is more controversy and a wider range of debate than is generally known.

## Figure 4.1. Scientists Views on Global Warming, 2003

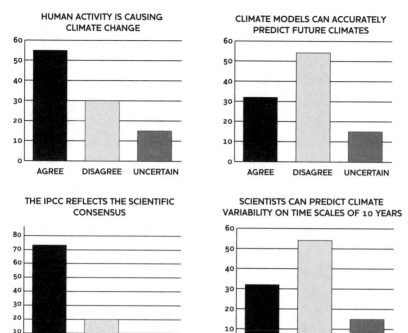

**Mr. Al Gore**

Al Gore depicts global warming as the crisis—the calling—of our time, as a moral imperative. As he boldly and memorably asks, is profit or the planet more important? Wouldn't a crisis of this scale, scope, and severity merit a colossal, global response? Yet his suggestions to change light bulbs, install weatherstripping, and drive less are vapid and anemic. They distract and deflect people from the real problem: an energy crisis that threatens the industrial and economic underpinnings and everyday lifestyles of billions of people. Gore's celebrated book, *An Inconvenient Truth*, recommends that the public "avoid overpackaged foods" and that we "don't stand in front of an open refrigerator door—leaving it open for just a few seconds wastes a lot of energy" (pages 182, 183). Okay, perhaps he is drawing the attention of people to the issue, rallying them to the cause, and fostering a grassroots movement that helps them believe they can make a contribution. Fine, but these suggestions only buy a little time.

Every U.S. presidential administration since 1973 should have wrestled with the issue, but didn't. The Clinton Administration represented 8 years of unforgivable procrastination and delay, particularly since you told us, Mr. Gore, that you were aware of the problem years earlier. For 8 years you and President Clinton occupied the White House, but no solution emerged. Not even a hint. Indeed, the Clinton Administration's policies made the situation worse by killing the fast-neutron-nuclear-reactor program.

So we need models, leaders, examples. I am reluctant to get personal, but I don't think Mr. Gore offers the world a valuable example of energy virtue. The average American household consumes approximately 10,500 kilowatt hours of electricity per year, but the Gore household consumes over 200,000 kilowatt hours annually, about 20 times the norm. Who leaves the greater "carbon footprint"? When leading the Indian Nation to independence, Gandhi collected salt, wore robes, and spun his own thread. Leading by example can be very powerful.

Personal energy habits not withstanding, critics will likely raise fewer objections if Mr. Gore and like-minded celebrities can convince the world to discontinue using the fossil fuels that create pollution and possibly alter the climate. Did Mr. Gore convince you or the world to kick fossil fuels, or just to change a few light bulbs?

Mr. Gore, I am certain we are running out of time, are you? I think we need an aggressive U.S. and global plan led by a true leader, do you? I don't think we will have much of a society or economy to worry about unless we immediately kick our fossil fuel addiction and commit to clean, renewable energy. You?

## MORE QUESTIONABLE EVIDENCE
## for GLOBAL WARMING

As with most things about energy, the data and numbers quantifying $CO_2$, other emissions, and their effect on the environment are all over the map and often highly qualified. "Information" often journeys from speculation to rumor to fact. Be careful.

### Greenland Ice Shelf

Some reports declare the Greenland Ice Shelf—about 2 miles thick and the largest mass of land ice in the northern hemisphere—is shrinking. Although P. J. Michaels believes the ice mass is largely in balance, much is being written about at least parts of Greenland experiencing significant warming. Some say that barley can again be grown in some regions. So what's new? Farmers grew barley there in the Middle Ages, likely explaining the region's name—Greenland. The *Economist* (September 9, 2007), reports dog sleds having problems in parts of Greenland, because the sleds are tricky to use when ice melts and the soil is mushy. The article implies that the melting is due to a rise in temperature of 1.5 degrees Celsius (2.7 degrees Fahrenheit) in the last 30 years. These claims beg a few questions. First, is global warming the certain cause of the melting? Second, since most of Greenland stays well below freezing the entire year, how much melting could a temperature rise of 1.5 degrees Celsius actually cause in Greenland?

Jonathan Gregory and his colleagues published a dramatic item in *Nature*, declaring that the climate of Greenland may warm enough during the next few centuries to eventually eliminate the Greenland ice sheet. The authors assert that the ice sheet contains enough frozen water to raise the world's sea level by about 23 feet, if the ice sheet melted completely. Likely?

What is going on in Greenland? Some recent studies indicate that over the past several decades Greenland's ice sheet is experiencing increasing summertime melting. Sounds ominous. However, a recent study concludes that Greenland's summer temperatures since 1940 are getting cooler. In 2004 Petr Chylek, of Los Alamos National

Laboratory, and his colleagues reported that Greenland experienced a rapid warming from 1920–1930, then steady temperatures to about 1940, followed by a cooling period lasting into the 1990s that lowered coastal temperatures by about 1 degree Celsius (1.8 degrees Fahrenheit) since 1940. Although coastal temperatures warmed modestly from 1992–1998, the general trend since 1940 is toward cooling temperatures. The U.S. Environmental Protection Agency, reporting on Chylek's research, declares that

> The cooling is even more pronounced further inland, at the summit of the ice sheet, where the summertime average temperature has declined at the rate of 3.9 degrees Fahrenheit per decade since records began in 1987.

Because the warming in the 1920s occurred before greenhouse gases began rising rapidly, the researchers conclude that the region's climate is naturally highly variable. They suggest that Greenland's temperature trends may be strongly influenced by the cyclic climate phenomenon known as the North Atlantic Oscillation. If they are correct, predictions of the ice sheet's fate would need to consider this relationship in conjunction with the assumed effects of human-induced climate change.

Writing in 2006, Chylek and colleagues report that Greenland's temperatures from 1905–1955 were warmer than its temperatures from 1955–2005. "We find no direct incidence to support the claims that the Greenland ice sheet is melting due to increased temperatures caused by increased atmospheric concentration of carbon dioxide. The rate of warming from 1995 to 2005 was in fact lower than the warming that occurred from 1920 to 1930."

Just when we think we've narrowed the argument, up pops another one. Dr. Ralph von Trese of Ohio State University, at a meeting of the American Geophysics Union on December 13, 2007, made a remarkable presentation. He believes Greenland ice is melting because of a thin "hot spot" in the earth's crust (below the ice shelf) in the northeast corner of Greenland. Iceland is similarly situated over a thin hot spot, which creates the island's volcanoes, hot springs, and geysers.

I admire the individuals and organizations that have such zeal for the issue that they go on-site to collect first-hand evidence of global warming. But let's state the obvious: Antarctica and Greenland are simply too big to poke around, snap some photos, and draw conclusions. Antarctica is 34 percent larger than Europe and 70 percent larger than Australia. Greenland is three times the size of Texas. One simply could not see enough on a visit to form a meaningful opinion. However, such select, happenstance observations could add to the confusion or misinformation. Hey, I've been to Miami, so I must know what the United States is like. Or I ate an apple, so I know what fruit tastes like.

## The Meltdown Showdown

Rising sea levels from melting ice? The melting of sea ice does not cause oceans to rise, because the ice is already in the ocean. Oceans rise only if significant melt water from landed ice and snow enter the ocean. In *An Inconvenient Truth*, Gore mentions one heck of a lot of melting:

- As measured by submarines, about 40 percent of the Arctic ice cap (sea ice) has melted.
- Icebergs are "calving" from landed ice and sliding into the sea.
- Satellite images of Greenland show that the ice cap has shrunk by half.
- Glaciers are melting in Kilimanjaro, Italy, Peru, the Himalayas, and Patagonia.

Should we worry about sea levels? According to the U.S. National Snow and Ice Data Center, if all the land ice in the world were to melt, then the oceans would rise approximately 230 feet. Further, if all of Greenland's ice were to melt, then oceans would rise about 23 feet. Similarly, if all of the Arctic's land ice were to melt, then oceans would rise about 23 feet. So let's let ocean levels end our speculation, since they can now give us the final answer to how much land ice has melted.

## Ocean Levels—The Final Arbiter

To determine the precise amount of land ice that has melted in the last 50 years, you only need a calculator, not a Ph.D. in oceanography. With prominent reports about glaciers disappearing, a 40 percent decrease in the thickness of some ice fields, and Antarctica and Greenland melting away at unprecedented rates, one would imagine the oceans have already risen substantially—but they haven't. So what's going on, and how much has really melted?

We know that when ice floating in water melts—whether ice cubes in a glass or icebergs in the ocean—the water level does not rise, because the ice already displaces water equal to its weight. However, ice melting on land will raise the level of the oceans. Some perspective: A rise of a few feet would swamp Florida and relocate the Atlantic coastline to someplace in South Carolina or Tennessee. Indeed, a lot of the world's land mass would be under water.

However, over the last 100 years the average temperature of the earth rose 0.6 degrees Celsius (1 degree Fahrenheit), and ocean levels over the last 50 years have risen only about 10 centimeters (4 inches). Thermal expansion caused one-half of the rise (2 inches), meaning the other 2 inches can be attributed to the melting of land ice. So, 2 inches is what percentage of the total possible rise in sea level? Okay, grab your calculator. Divide 2 by 12 to convert inches to feet.

> Less than one-tenth of one percent of the world's land ice has melted in the last 50 years.

Now divide by 230 the amount in feet that oceans would rise if all ice melted. The result shows us that only 0.07 percent melted. *That means that less than one-tenth of one percent of the world's land ice has melted in the last 50 years.* This is a far cry from reports that declare that 10, 20, and even 40 percent of major ice fields are melting. One-tenth of one percent sure doesn't square with all of the alleged melting that many proclaim. If they are right about the melting, then there is a lot of freezing going on at the same time. More perspective: If 1 percent had melted, then the oceans would have risen approximately 14 inches. This analysis is simple and absolute—discussion over.

Professor P. Winsor bluntly declares "there was no trend towards a thinning ice cover during the 1990s" in the Arctic Ocean.

> The ice cover of the Arctic Ocean is considered to be a sensitive indicator of global climate change. Recent research, using submarine-based observations, suggests that the Arctic ice cover was thinner in the 1990s compared to an earlier period (1958–1979), and that it continued to decrease in thickness in the 1990s. Here I analyze subsurface ice thickness (draft) of Arctic sea ice from six submarine cruises from 1991 to 1997. This extensive data set shows that there was *no trend* towards a thinning ice cover during the 1990s. Data from the North Pole shows a slight increase in mean ice thickness, whereas the Beaufort Sea shows a small decrease, none of which are significant. Transects between the two areas from 76 degrees N to 90 degrees N also show near constant ice thicknesses, with a general spatial decrease from the Pole towards the Beaufort Sea. Combining the present results with those of an earlier study, I conclude that the mean ice thickness has remained on a near-constant level around the North Pole from 1986 to 1997 (emphasis added).

### Bare Facts About Bears

Many presentations on global warming show forlorn polar bears stranded on an isolated iceberg, threatened with drowning or starvation. Al Gore's frightened bear tries to climb atop floating ice that breaks up as the bear jumps aboard. School children, soccer moms, and Citizen Joes have rallied to their aid. The poor, beleaguered polar bear has become the *poster child* of global warming. Though stirring and compelling, this image tells us nothing.

The following is summarized from a IUCU/SSC Polar Bear Specialty presentation in 2005. According to the report, the number of bears in about one-third of the Arctic is unknown. In another third, where the estimated bear population is about 9000, it is unknown whether the bear population is increasing or decreasing. In the final third, scientists conclude that one population of about 5500 bears will *likely* decrease, because over 75 percent of computer simulations project population decreases. However, a group of about 7000 bears, also in this final one-third of the Arctic, is described by scientists as *likely* to increase, because over 75 percent of computer simulations predict polar bear population will grow. Therefore, I conclude from the same data most often used to assert the decreasing numbers of polar bears that the population of polar bears appears fairly stable.

Polar bears are difficult to count. They can move around an area of 100,000 square miles over their lifetime. The total number of polar bears worldwide is conventionally estimated at 20–25 thousand, although some references put this range at 16–40 thousand. If polar bears travel so far, and the population estimates are so widely variable, then how can one count them or possibly declare for a fact that the population is decreasing or increasing?

Besides, polar bears have been around for over 250,000 years. They have survived ice ages and interglacial periods when there was little or no ice.

We *should worry* about threats to bears, other animals, plants, and humans from the increasing use of fossil fuels. How about the proposed tearing up of Alaska, threats to the ANWR, NPRA, and the Beaufort Sea, and leaky oil pipelines?

## The Tale of Melting Glaciers Is All Wet

If the sad stories of drowning polar bears don't grip you, then the woeful tales of glaciers rapidly melting away due to rising temperatures of global warming may. Global warming threatens cuddly bears and also destroys picturesque vistas and exciting skiing and hiking locales that humans enjoy. Now *that* hits close to home. Or does it?

Are *some* glaciers or *all* glaciers melting? There are over 100,000 glaciers in the world, covering 10 percent of the world's land area—an area about the size of South America. Antarctica is home to 80 percent of the world's glaciers; about 12 percent are in Greenland. These glaciers contain about 75 percent of the world's fresh water. The melting of 10, 100, or 1000 glaciers proves nothing unless we know the condition of the rest of them. Despite the hand wringing and hyperbole, many glaciers are growing.

Postcard-pretty pictures of melting glaciers and lone polar bears bring nothing to the discussion on global warming except emotion and, I think, unnecessary confusion.

## Weather Threats?

According to Al Gore and others, global warming is causing more frequent and more violent hurricanes, tornados, drought, and more. Are these threats verifiable, proven by data? Or are these potential threats examples of exaggeration that increase readers and viewers?

One of the world's foremost experts on hurricanes—the man who pioneered the practice of predicting a season's worth of hurricane activity months in advance and whose hurricane forecasts have been used by insurance companies since 1983 to set premiums—was interviewed in 2005 by Kathy Svitil for *Discover* magazine. Dr. William M. Gray said,

> The Atlantic has had more of these storms [hurricanes] in the last 10 years or so, but in other ocean basins, activity is slightly down. Why would that be so if this is [due to] climate change? The Atlantic is a special basin. The number of major storms in the Atlantic also went down from the middle 1960s to the '90s, when greenhouse gases were going up.

"Nearly all of my colleagues who have been around 40 or 50 years are skeptical as hell about this whole global warming thing. But no one asks us."

Hurricane Katrina was a devastating tragedy. It hit a major city and population center with terrifying force. The media could easily and vividly cover the event. The media coverage amplified and magnified our attention. But one historic storm does not make a trend. For U.S. hurricanes by decade, see Figure 4.2.

Here is a longer excerpt from Dr. Gray's interview:

**Q:** You don't believe global warming is causing climate change?

**A:** No. If it is, it is causing such a small part that it is negligible. I'm not disputing that there has been global warming. There was a lot of global warming in the 1930's and '40's, and there has been warming since the middle '70's, and especially in the last 10 years. But this is natural, due to ocean circulation changes and other factors. It is not human induced.

**Q:** That must be a controversial position among hurricane researchers?

**A:** Nearly all of my colleagues who have been around 40 or 50 years are skeptical as hell about this whole global warming thing. But no one asks us. If you don't know anything about how the atmosphere functions, you will of course say "Well, greenhouse gases are going up, the globe is warming, they must be related." Well, just because there are two associations, changing with the same sign, doesn't mean that one is causing the other.

**Q:** Why is there scientific support for the idea?

**A:** So many people have a vested interest in this global warming thing—all these big labs and research and stuff. The idea is to frighten the public, to get money to study it more. Now that the cold war is over, we have to generate a common enemy to support science, and what better common enemy for the globe than greenhouse gases?

Dr. Gray's statement "But no one asks us" is quite typical. Many scientists I interviewed for this book told me that "politics" was the most formidable barrier to fixing the nation's energy problems. They told me that scientists essentially have "no voice." One even said "We're not invited to the table except to say what we are told to say, [because] funding is the issue." Policy and special interests over good science are sure ways to perpetuate bad policies and political ignorance.

I and others were greatly disturbed when the Clinton Administration appointed Mr. Richardson, *an attorney*, to be the Secretary of Energy. Couldn't they find somebody more qualified who actually knew something about energy? I questioned whether Mr. Richardson knew anything about the basics of energy science or the energy industry. One critic wondered whether Richardson knew the difference between a BTU and a BLT. It's like having an engineer give advice on legal matters. It's beyond senseless.

## Figure 4.2. U.S. Hurricane Strikes by Decade

| DECADE | SAFFIR-SIMPSON CATEGORY* | | | | | ALL 1, 2, 3, 4, 5 | MAJOR 3, 4, 5 |
|---|---|---|---|---|---|---|---|
| | 1 | 2 | 3 | 4 | 5 | | |
| 1851–1860 | 8 | 5 | 5 | 1 | 0 | 19 | 6 |
| 1861–1870 | 8 | 6 | 1 | 0 | 0 | 15 | 1 |
| 1871–1880 | 7 | 6 | 7 | 0 | 0 | 20 | 7 |
| 1881–1890 | 8 | 9 | 4 | 1 | 0 | 22 | 5 |
| 1891–1900 | 8 | 5 | 5 | 3 | 0 | 21 | 8 |
| 1901–1910 | 10 | 4 | 4 | 0 | 0 | 18 | 4 |
| 1911–1920 | 10 | 4 | 4 | 3 | 0 | 21 | 7 |
| 1921–1930 | 5 | 3 | 3 | 2 | 0 | 13 | 5 |
| 1931–1940 | 4 | 7 | 6 | 1 | 1 | 19 | 8 |
| 1941–1950 | 8 | 6 | 9 | 1 | 0 | 24 | 10 |
| 1951–1960 | 8 | 1 | 5 | 3 | 0 | 17 | 8 |
| 1961–1970 | 3 | 5 | 4 | 1 | 1 | 14 | 6 |
| 1971–1980 | 6 | 2 | 4 | 0 | 0 | 12 | 4 |
| 1981–1990 | 9 | 1 | 4 | 1 | 0 | 15 | 5 |
| 1991–2000 | 3 | 6 | 4 | 0 | 1 | 14 | 5 |
| 2001–2004 | 4 | 2 | 2 | 1 | 0 | 9 | 3 |
| 1851–2004 | 109 | 72 | 71 | 18 | 3 | 273 | 92 |
| Average Per Decade | 7.1 | 4.7 | 4.6 | 1.2 | 0.2 | 17.7 | 6.0 |

Number of hurricanes by Saffir-Simpson Category to strike the U.S. mainland each decade.

*Only the highest Saffir-Simpson Category to affect the United States has been used.
Year 2006 had only one-third the number of hurricanes compared to 2005.
Ocean surface temperature in the Atlantic in 2006 was 1.8°F cooler than in 2005.
Sources: National Weather Service, National Hurricane Center.

## THE DIFFICULTIES and UNCERTAINTIES of MODELING GLOBAL CLIMATE

The climate of the earth is extremely difficult and perhaps impossible to model. Many complex, overlapping, and compensating mechanisms interact. The atmosphere, clouds, the earth, volcanic and tectonic activity, the oceans, ocean currents, storm systems, rainfall, ice, all plant life, all animals, sunspots, solar radiation, the earth's tilt, and numerous other variables interact to affect global climate in complex ways. As scientists learn more about elements of this magnificent global system, theories about the system as a whole and about its elements change constantly. Some effects are negative, others positive, some neutral, yet others are cyclical, modulating, or self-compensating. Then there are threshold phenomena, such as the often talked about "tipping point" or "point of no return," that add another layer of complexity.

One reason for establishing the Intergovernmental Panel on Climate Change (IPCC) was to create a widespread scientific basis and consensus on global climate. That consensus has not yet emerged. Yes, humans affect global climate, but details are hotly debated. Using IPCC to determine policy seems very dangerous.

Billions of dollars have been spent to create such models, but several published critiques argue that no consistently reliable model has yet been created.

> ❯ Christopher Monckton, an advisor to Margaret Thatcher and one of Britain's leading public intellectuals, wrote a pointed article for London's *Sunday Telegraph* that strongly criticizes prevailing notions of global warming, especially apocalyptic visions. To supplement the article, the *Sunday Telegraph* posted Monckton's calculations and references. The supplement—40 pages long and containing almost 90 references, including works from many respected climate scientists—is entitled "Apocalypse Cancelled." A worthy read. Monckton thoroughly examines the topic of global warming, including the assumptions the IPCC used in

building its climate model, from a perspective strongly rooted in scientific fundamentals. In contrast, Mr. Gore's approach appears more political and more emotional. Monckton has challenged Mr. Gore to a debate. Wouldn't that be fascinating?

- On April 6, 2006, 60 climate scientists wrote a letter to the Canadian Prime Minister calling for public consultation to examine the government's climate-change plans and to review the Kyoto Protocol.

- Over 15,000 articles were published every month in 2007 with the key expression "global warming" in the title or the text. Most depict frightening future scenarios, but many are still skeptical.

Regardless of what you believe or how confused you are about global warming, don't worry because it doesn't matter. It would be insane to bet that global warming is not caused principally from industry and the burning of fossil fuels, because if you lose that bet, the consequences are simply too severe. So don't make the bet. Besides, surer bets await us: The United States and the world have the technologies and the wealth to essentially eliminate human-caused $CO_2$ and other human-generated greenhouse gases to acceptable levels *and* to solve other serious environmental and economic problems at the same time. We face, however, a race against time (see Chapter Sixteen).

> The climate of the earth is extremely difficult and perhaps impossible to model. Many complex, overlapping, and compensating mechanisms interact.

Let's consider some potentially valuable and not-so-valuable solutions.

## SOLUTIONS?
### Conservation Is No Remedy

Many people, particularly after seeing Gore's movie, believe that conservation is *the* answer. However, upon brief reflection, this proposed solution collapses. Conservation is simply no answer. Discontinuing differs from conserving, which is wholly insufficient to the scale, scope, and severity of the crisis we confront. Changing people's energy habits will be difficult, especially the energy habits of the privileged. Some must lead. Some must set examples. Some commendable people will conserve energy by changing their habits voluntarily, but conservation alone buys only a little time before a sufficiently robust solution arrives. As a result, conservation makes only a minor, modest contribution to a solution. Let's do the numbers.

If every automobile in the world used 20 percent less fuel, then the world would still run out of oil in the near, foreseeable future, and pollution, while reduced, would still add unacceptable amounts of $CO_2$ and other pollutants to the air and water every year. Yet so what if each vehicle consumes less gasoline, if at the same time population growth means many more cars on the world's roads? In 30 years 2–3 billion more people will inhabit the planet, and people in developing countries are just beginning their love affair with the automobile. What does a hypothetical reduction of 20 percent in fuel use yield the world? It would save about 6 billion barrels of oil per year, but in about 9 years the world would still be using more than we are today. In the grand scheme of things, the savings would buy the world about 7 more years before running out. Somewhat helpful, but no solution. And since we all know we cannot instantly or easily increase gas mileage by 20 percent using internal combustion engines, the answer is clear: We must change our automotive propulsion systems and quit using fossil fuels. See Chapter Fourteen for details.

> Conservation makes only a minor, modest contribution to a solution.

## Feeble Thinking Is No Remedy

Some authors offer no suggestions at all. Mr. Gore does, but his suggestions are anemic and disappointing. They delay disaster for a few years, but offer no chance of solving the energy problems of the United States or the world. At best, his suggestions return us to the pollution rates of 1970. Are 1970 levels okay with him? Again, somewhat helpful, but no solution. Let's consider a few of his proposals.

- **Efficient appliances.** Efficient appliances will cure climate change and the energy crisis? Laughable. We've already accomplished a lot here. Deployment of further improvements will take a long time.

- **Higher-mileage cars.** Duh. Why was the Corporate Average Fuel Economy (CAFE) law not enforced during the Clinton Administration? The solution is to switch to plug-in hybrids and all-electric cars.

- **Other transportation efficiencies.** Walk? Bike? Bus? Light rail? Great for many purposes, but not for getting the groceries home, the kids to soccer practice, and bulk goods across the continent.

- **Renewable energy sources.** Of course. Any fifth-grader knows this. But which sources? Trade offs among sources? Opportunity costs? Funding? Timing?

- **Capture carbon from power plants and sequester it in the ground or in the ocean.** Not yet. There are issues of long-term safety, so the time for this procedure has not yet arrived—it may never. Where in the ground? The concentration of carbon dioxide in the ocean is already too high. How many power plants are conveniently located next to a large underground reservoir? How about the long-term safety of such a scheme? What if this storage of carbon dioxide "burps" (releases large amounts of carbon dioxide into the atmosphere in a short time)? Depending on the concentration, it could kill everybody in the area. How long would this carbon dioxide have to be sequestered? Not 100 or 1000 or 100,000 years, but *forever.*

◗ **Capture greenhouse gases and other nasty pollutants at the plant.** This is a good idea, but it is expensive and difficult. While helpful, it is not technically or economically feasible to capture enough. Besides, how about breakdowns and cheating? Another frequent tactic is to contribute large sums of money to political campaigns to secure favorable legislation. A solution for some, and a sure disaster for the rest of us.

◗ **Gasify coal.** If we continue with business as usual, we will probably be forced to gasify coal for many uses, and convert coal to oil to satisfy our transportation needs. Then, instead of our coal reserves lasting 230 years, they would last no more than 100 years if we are forced to convert coal to replace depleting oil and natural gas.

◗ **Burn more natural gas instead of coal.** We do quite a lot of this now. If we use more of our natural gas to generate electricity, then we will run out long before the middle of this century. Also, this option is not much of a bargain in regard to carbon dioxide. Recall that every pound of natural gas burned creates 2.75 pounds of carbon dioxide.

## Does Planting Trees Buy Us "Forgiveness"?

Planting trees helps compensate for (offset) greenhouse gas emissions, because trees absorb carbon dioxide and release oxygen. Alas, many forestry-offset projects were conceived or conducted in ways vulnerable to criticism, drawing the projects' net benefits into question. Significant concern also arises over the permanence of carbon storage in trees and forests, since future clearing of the forests would return the stored carbon to the atmosphere. In some instances, foresters have planted many trees of the same variety together, thereby creating a "monoculture" environment, which actually destroys the biodiversity necessary for a healthy forest. Unintended consequences?

In Ecuador, the Dutch FACE Foundation operates a carbon-offset project in the Andean Páramo area. The project involves the planting of 22,000 hectares with eucalyptus and pine, of which 20,000 hectares were certified under the Forest Stewardship Council system. Following an investigation, the non-governmental organization Acción Ecológica criticized the project in May 2005 for destroying a valuable Páramo ecosystem by introducing exotic tree species, causing the release of much soil carbon into the atmosphere, and harming local communities that had entered into contracts with the FACE Foundation to plant the trees. Who knew planting trees could be so involved and controversial?

How many trees would we need to plant? A typical tree planted in the tropics removes about 50 pounds of carbon dioxide from the atmosphere each year. Sounds good, but what does that mean? Each average tree absorbs the $CO_2$ generated by burning 3 gallons of gasoline. So, one would have to plant about 250 trees to offset one year of driving the average U.S. car an average annual distance. Trees in the rainy tropics grow 3 times faster than trees in temperate zones. By the way, the Gores would have to plant tens of thousands of trees in a temperate zone to compensate for the energy the Gores use.

However, when trees die, their carbon goes back into the environment, for a net reduction of zero. So what exactly does planting trees do for us? *Nothing*. Tree-planting-as-salvation is based on weird science practiced by some politicians and their uninformed disciples.

Indeed, "Planting Trees to Save the Planet is Pointless," announces the *Guardian* [U.K.] in the headline to an article by Alok Jha. The article quotes

> Each average tree absorbs the $CO_2$ generated by burning 3 gallons of gasoline.

Ken Caldeira, co-author of a study released by the Carnegie Institute:

> The idea that you can go out and plant a tree and help reverse global warming is an appealing, feel-good thing....

To plant forests to mitigate climate change outside of the tropics is a waste of time.

Jha writes, "Prof. Caldeira said planting trees was a diversion, letting consumers pollute more. He said it would be better to transform the way energy was derived and used, for instance through investment in renewable and carbon-free electricity generation."

In a separate interview reported by R. Butler, Caldeira said, "in terms of climate change, we should focus our efforts on things that can really make a difference, like energy efficiency and developing new sources of clean energy." The report, titled "Climate Effects of Global Land Cover Change," was co-authored by S. Gibbard, K. Caldeira, and others.

> Tree-planting-as-salvation is based on weird science.

Did you get that, everybody? We must transform the way energy is derived, developed, and used; we must *transform our energy system*. That is the point of this book. Read on for details of costs, timetables, and implementation.

## Carbon Credits, Carbon Offsets, and Cap-and-Trade Schemes are Not Remedies

The Gores did purchase carbon credits to offset their polluting lifestyles. Mark Steyn, in an opinion essay in the March 4, 2007, edition of many national newspapers, reports that the Gores purchased the carbon credits from a company owned at least partially by Mr. Gore. Many critics believe purchasing carbon credits makes the rich feel good, but does nothing to reduce the *net* output of $CO_2$. Many are blunter. They believe, and I agree, that purchasing carbon credits is a sham. Has anyone coined the term *eco-fraud*?

> We must transform the way energy is derived and used; we must transform our energy system. That is the point of this book. Read on for details of costs, timetables, and implementation.

I would really like to see a well-quantified, reputable, non-self-interested report illustrating the costs and detailing who benefits, who gets the money, who administers the programs, who brokers the transactions, and how the practice reduces $CO_2$ emissions. Do these "offset companies" use the money to build non-polluting energy systems? Would these systems have otherwise been built? Do wealthy people purchase carbon credits from poor people because the poor can't afford to use more energy? How do poor people purchase carbon credits? Anybody seen such a report?

> We must transform the way energy is derived and used; we must transform our energy system.

The designers of carbon-offset and carbon-credit programs intend to bring the world closer to carbon neutrality. The programs are complicated, and many don't think they achieve the desired ends. Even some of those trying to explain the programs admit they don't get it. Consider this response, appearing in *The Vail* [Colorado] *Daily*, to a reader's question:

> I've heard renewable-energy credits, which are supposed to help create more renewable power, aren't good. I buy wind power for my home. Am I buying lies?
>
> —Ryan

The partial response:

> Renewable-energy credits are quite the sticky bun these days, Ryan. Let's first take a step back to explain them and their close cousins, carbon offsets. But I must warn you, when we're done you still may not understand credits and offsets... Until and unless the U.S. Environmental Protection

> George Monbiot has compared carbon offsets to the practice of purchasing indulgences during the Middle Ages.

Agency regulates carbon dioxide and other greenhouse emissions, there is no regulatory body to standardize, enforce or simplify these commodity markets. Thus, buyers beware. Does the credit money you just spent really go to a new renewable project? How much? Why does one company cost $10 per carbon offset and other $6? And you thought paper versus plastic was tough… Then there's the issue of 'retail therapy.' Will people feel entitled to drive more, heat driveways and leave lights on if they can just buy their way out of consumption sin? There is concern that credits and offsets will cause more energy consumption when what is needed is greater conservation. Credits and offsets are ideal only for energy consumption we cannot currently avoid. Right now, the annual growth in energy consumption alone in the U.S. is greater than all the renewable power generated every year… So all clear now, Ryan?

— Energetically, Terra.

Mr. Gore, did you know that the provisions for carbon credits and carbon offsets written into the Kyoto Protocol were intended to provide a mechanism for *carbon neutrality* for companies and industries that simply *cannot* lower their emissions because of the nature of their processes. Carbon credits and carbon offsets were not intended to provide a means for wealthy individuals to repent for their consumption sins. Examples of target industries include manufacturers of cement, steel, textiles, and fertilizers. In these cases carbon credits and offsets might make some sense, but let me repeat our theme song: *The only real solution is to kick the fossil-fuel addiction.*

Has anyone coined the term *eco-fraud*?

### Sacrilege or Salvation?

The option to purchase offsets raises other questions—some practical, some political, some economic, some ethical. Do offsets legitimize the continued production of greenhouse gases? Do offsets encourage polluters to pollute more? Do these markets actually reduce carbon emissions? The measurement of the true harm of carbon emissions and the benefits of offsets are exceedingly difficult.

Let's consider an analogy. What would be the implications of an offset program by which someone who wants to betray his or her spouse could make a payment to someone who pledges to be faithful? A group of activists in Machynlleth, Wales, did just this by setting up a spoof website <www.cheatneutral.com> to illustrate a fundamental contradiction in offsetting practices. Such programs make it seem acceptable to continue to maintain the status quo, even if immoral or harmful. The continued generation of pollutants in an era of global environmental change is certainly harmful. I'll leave it to others to draw the moral conclusion.

George Monbiot, an English environmentalist and writer, has compared carbon offsets to the practice of purchasing indulgences during the Middle Ages. This practice allowed people to purchase forgiveness for their sins, rather than to repent or stop sinning. Monbiot also says that carbon offsets are an excuse for business-as-usual pollution.

### Boon or Bane?

In 2007, Fiona Harvey and Stephen Fidler of the *Financial Times* investigated the carbon-offsets industry. Among their findings, quoted below:

- Widespread instances of people and organizations buying worthless credits that do not yield any reductions in carbon emissions.
- Industrial companies profiting from doing very little—or from gaining carbon credits on the basis of efficiency gains from which they have already benefited substantially.

◗ Brokers providing services of questionable or no value.

◗ A shortage of verification, making it difficult for buyers to assess the true value of carbon credits.

### The Nitty Gritty

Offsets and carbon credits represent an accounting nightmare. Their virtues and effectiveness are uncertain. They look good and feel good, but do no good. They should be abandoned. A straight carbon tax would be easier to administer and would surely be more equitable. It also would likely achieve the intended result.

No reduction or carbon-neutral program will save us. While they can buy us some time, they can also deaden our sense of urgency.

### Cap-and-Trade Programs

Cap-and-trade programs are cousins of carbon credits and carbon offsets. They have been successful in reducing other pollutants, but here again they only *reduce* emissions when we actually need to *eliminate* them. Some industries cannot avoid producing carbon dioxide as part of their manufacturing process, and we must do everything possible to help them become more carbon efficient. When all of our electricity is produced by clean, renewable sources, then we can tolerate the unavoidable emissions by some industries. That would be the ultimate offset program.

## A REAL SOLUTION

I wish Al Gore would use his unprecedented platform to talk more about the benefits of renewable fuels and the prospect of energy independence. Mr. Gore must certainly know that the United States must develop and deploy clean energy, retire all fossil-fuel-burning power plants, and curtail the use of oil. *I especially wish Mr. Gore would have mentioned nuclear energy.* To my knowledge he said nothing about fast neutron reactors or the proposed Global Nuclear Energy Partnership

(GNEP). *These are unforgivable omissions.* Too much political risk? Too much misinformation and ignorance surrounding nuclear energy, although it already produces a full 80 percent of U.S. carbon-free electricity?

Nuclear energy represents the *only* total solution (see chapters Ten and Eleven). The people of the United States and the world will not be able to conserve their way out of the energy problem. Nor can wind, solar, or hydrogen fill the gap in the time needed. As solar cells become more efficient, they may become the dominant source of energy 100 years from now—maybe. The only practical and affordable solution for today is to aggressively deploy nuclear power plants now—particularly fast neutron reactors.

Although solving the energy problems of the world and the United States benefits just about everyone, an undeniably inconvenient truth is that the problems of energy transition and global warming simply cannot be solved without causing hardship to some very important industries, such as the oil, coal, and automotive industries—all with

> The United States and the world will not be able to conserve their way out of the energy problem.

strong lobbies—and to people who work in and near these industries. Some people will be displaced. Fuel distribution and other automotive services must also be greatly modified to accommodate any new transportation system.

Governments and private industries could manage displaced firms and people. They can minimize the harms while realizing earth-saving benefits. People throughout history have suffered hardship, displacement, and inconvenience in their quest for freedom, particularly in wars of freedom and independence. This "war" is as serious and challenging as any we have fought in the past. The assault on our way of life from pollution is as serious as any war ever fought. The one big difference is that this time the whole world should be on the same side. Our common enemies are the burning of fossil fuels and the pollution they cause. Here's another chorus from our theme song: *This massive global problem needs a huge, global solution.*

The solution is not mysterious. In fact, it is quite simple, because there is only one workable, effective solution: fast neutron nuclear reactors, the Global Nuclear Energy Partnership, and a rapid transition in our transportation fleet to hybrid plug-ins and all-electric vehicles. The cost? Very affordable, as we shall see.

## TRUE BELIEVERS, CONTRARIANS, DISMAYED SKEPTICS, and the DISINTERESTED

I have done my best to lead you through official statements, thoughtful opinions, and blatant propaganda surrounding the debates on global warming. Aspects of these debates now seem to hold permanent, prominent positions in media broadcasts and publications. True believers and contrarians alike passionately hold their views. Many others in the public at large seem firmly confused or disinterested. While I am a skeptic, I am deeply dismayed by the fear and furor surrounding these unnecessary debates. Let me repeat my gospel: Fossil-fuel use, not global warming, is the real problem.

The medieval writer Dante supposed that the gate to hell was capped by a foreboding sign: "Abandon hope all ye who enter here." I think the same warning applies to those who heedlessly enter a future dependent on fossil fuels. Why abandon hope when we can simply abandon coal, oil, and natural gas? Since we must forsake fossil fuels sometime in the near future anyway as they become depleted, why not get off fossil fuels earlier than later and save the world trillions of dollars?

> There is only one workable, effective solution: fast neutron nuclear reactors, the Global Nuclear Energy Partnership, and a rapid transition in our transportation fleet to hybrid plug-ins and all-electric vehicles.

The best way to begin is to do everything we can politically to get the necessary changes started. What can we expect? If the best way to tell the future is to peer into the past, then we can reasonably expect our political leaders to do little to address, much less to solve, this mammoth problem. In the United States, no one can or will wrestle with Social Security issues or healthcare, and we are even worse off on energy matters. The public must take action, take control, stay in leaders' faces, and insist on specific, affordable, timely, effective policies designed to hasten the transition to renewable energy sources.

The ballot box is one of the public's most powerful tools, but elected officials and vested interests also possess formidable tools: cash, campaign contributions, and lobbying skills. Also, the world has little time. Those interested in environmental health and energy must convince each other, fellow citizens, our political and corporate leaders, and the world to act now. A policy that insists on the deployment of all-renewable energy sources and the rapid replacement of fossil fuels is the only sane course. Only a fool would resist switching completely to renewable, clean energy sources as fast as possible.

## BOTTOM LINE

- The argument on global warming is meaningless, is unresolvable, and doesn't make any difference unless it becomes the motivation for abandoning fossil fuels.
- Don't let all of the discussion about global warming distract you. Only total abandonment of fossil fuels will do.
- Without a cry to abandon fossil fuels and adopt the inevitable nuclear solution, Mr. Gore's message misses the mark.
- Offsets, credits, and cap-and-trade schemes will *never* solve anything. They may buy us a little time and a little denial.

- Cap and Trade programs have not reduced emissions significantly in Europe nor will they make much of a difference in our quest to energy independence. This effort would be better spent supporting a more aggressive transition from fossil fuels to clean renewable energy sources. Further, this is another scheme ripe for abuses, and accounting shenanigans.

- Whatever your position on global warming, don't bet against the dire consequences of global warming. You don't have to.

- We must take action immediately. Money will not be the problem in the transition to clean, renewable energy. Time and political leadership will be. On the political side I remain hopeful, because in times of need a good national leader often emerges.

- Mr. Gore, I really wish you would have reached across the aisle in a spirit of cooperation to solve this most serious problem. A Nobel Prize, yes. A place in the Energy Hall of Fame, not yet. But you can still get in. I stand by to help in any way I can.

*Chapter 5*

# ENERGY PROBLEMS AROUND THE WORLD

## NO RELIEF IN SIGHT

Let's take a look around the world. Keep in mind that the depletion of energy resources and mounting fossil fuel pollution are not national problems—they are global.

### UNITED STATES

The U.S. population reached 300 million in October of 2006. At the present growth rate, the U.S. population may reach approximately 390 million in 30 years. The U.S. population may hit 470 million in 50 years. If you think we have congestion and polluted air now, then imagine how bad it will be if we keep doing the same dumb things.

U.S. oil reserves are 22 billion barrels, and the consumption rate is 7.5 billion barrels per year. With no population growth, no improvement in standards of living, no imports of oil, and no change in the consumption rate, these oil reserves would last only 3 years. If that

97

doesn't give you heartburn, then try this—the rest of the industrial world is in the same boat.

Under the same assumptions, U.S. natural gas reserves would last 9 years. Coal, by far the most abundant fossil fuel in the United States, would last for over 220 years at our present consumption rate, but world supplies will last only 175 years. When burned, coal puts a host of nasty stuff into the environment. As we burn more deeply into our reserves, we will burn dirtier coal containing more and more sulfur and other foul materials. Think we have ample coal reserves? Think again. If nothing is done now to create alternative, renewable energy sources, then we will be forced to use coal to produce oil and gasoline, since we will run out of these resources long before we will run out of coal. And if we do that, then our coal reserves will disappear in less than a century. If we use coal to generate electricity and also divert some coal to make gasoline, then the coal would last only about 80 years, if the conversion was 100 percent efficient, which it isn't. At 50 percent conversion efficiency, which is more likely, the coal would last about 50 years, then it too would be gone. Just when we think we have found a solution, we uncover other problems.

Say it with me: The only real solution is to quit burning fossil fuels.

## EUROPEAN UNION

*Europe consumes on a per capita basis about one-half the energy and fuel consumed in the United States.* This is pretty much across the board— oil, coal, and natural gas. The European Union's (EU) gross domestic product (GDP) is about the same as the United States'. The EU population of 460 million people will likely shrink in the next 30 years.

With 7 percent of the world's population, the EU consumes nearly 18 percent of the world's energy. They have oil reserves of 7.3 billion barrels and consume 5.5 billion barrels per year. Relying on their own oil reserves, EU oil would last less than 2 years. Their domestic natural gas, at present consumption rates, would last about 7 years. EU coal would last 44 years.

## CHINA

China, for all its industrial strength, may ultimately be deprived the reward of a more prosperous life for its people—one they so richly deserve—if their energy problems are not resolved.

China has about 4.5 times as many people as the United States, and they live on about the same land area. China, a rapidly growing world economic power, will have ever-increasing needs for energy for three reasons:

> **"China has become the world's factory, but also its smokestack."**

- Rapidly growing population—about 1.3 billion in 2008 will become about 1.6 billion in 2038. That is, in the next 30 years China's population will increase by 300 million people, the current size of the total U.S. population.
- Growing appetite for the good life.
- Rapidly expanding economy—China's economic growth rate in 2005 was about 10 percent.

Consider this report from the *New York Times* (December 21, 2007). The title of the article says much: "China Grabs West's Smoke-Spewing Factories." According to authors Joseph Kahn and Mark Landler, China is buying and dismantling piece by piece dozens of large factories from Europe, then shipping and reassembling those factories in China.

> **China in 2007, for the third year in a row, increased its power output by more than the total capacity of Britain.**

In its rush to re-create the industrial revolution that made the West rich, China has absorbed most of the major industries that once made the West dirty. Spurred by strong state support, Chinese companies have become the dominant makers of steel, coke, aluminum, cement, chemicals, leather, paper, and other goods that faced high costs, including tougher environmental rules, in other parts of the world. China has become the world's factory, but also its smokestack.

Such industries consume mountains of fossil fuels and belch black toxic clouds. Does China have the resources?

## Oil

China has 18.3 billion barrels of proven reserves of oil, yet the Chinese consume 2.3 billion barrels per year. So, with no population growth, no improvement in standards of living, and no importation of oil—that is, if the Chinese rely entirely on their own oil resources—their oil would last less than 8 years to bone dry. All gone. This fact must have the Central Committee downing lots of Mao-Tais, one of their most respected liquors. China plans to meet future oil demand in part from OPEC—but so does everybody else. Good luck. China has 550 million people fit for military service. Such military clout will make a large statement if the Chinese go after energy reserves in other parts of the world.

## Natural Gas

The Chinese are in better shape when it comes to natural gas. According to the 2004 edition of the CIA *World Fact Book*, China's proven reserves are 90 trillion cubic feet, and consumption is 1.2 trillion cubic feet per year. With no population growth, no improvement in lifestyles, and no imports, these reserves would last approximately 75 years to bone dry. Not too bad, but this is only one lifetime—back to the Mao-Tais.

## Coal

China is a coal heavyweight, and the Chinese use a heck of a lot of it. I can personally attest to this. In 1985, during my first visit to Beijing, people were wearing face masks to combat the dirty air. I understand this is still the practice in parts of China. Chinese coal reserves total 126 billion tons, and consumption is about 1.5 billion tons per year. These reserves should last about 85 years with no population growth, etc. However, these reserves don't seem to change from year to year, leading some statisticians to believe the stated reserves are likely inaccurate. In addition, most of the coal, primarily used to generate electricity, is relatively low-grade, generating considerable pollution. In some places in China, while your eyes burn, it is difficult to see across the street.

## Hydropower

The world must congratulate China on its development of vast hydro-electric resources. Over the last several years the Chinese have had under construction 35,000 megawatts of electrical capacity, about 16 percent of their total electricity needs. Amazing. This is a huge hydro-electric undertaking, as large as the next 3 largest programs (Brazil, India, and Iran) combined. According to the World Energy Council, this is enough capacity to supply the electrical needs of 30–50 *million* people in China. The largest project, the Three Gorges System on the Yangtze River (18,200 megawatts), is scheduled to be completed and totally online by 2009.

The Three Gorges dam project costs approximately $25 billion, about $1400 per installed kilowatt of generating capacity. This figure is about the same as for nuclear-generated electricity ($1500 per installed kilowatt). Experts project that China has the potential to produce 290,000 megawatts of economically feasible hydropower. China currently has approximately 50,000 megawatts of hydropower online. For perspective, 50,000 megawatts is equivalent to almost 85,000 relatively large (1.65 megawatt) windmills, and even then you would need a suitable energy-storage system. So many windmills are required because hydropower is 24/7, but suitable wind blows only about 35 percent of the time. If China develops all of its hydropower potential, and if the Chinese grid could accommodate all of this hydropower, then hydropower could replace all of the coal-burning and other electricity-generating power plants in China, at present consumption rates. However, even this staggering addition of electrical power would not be adequate for China's anticipated growth in population and economic development. Any shortfall could be made up with nuclear energy.

Hydropower from dams is subject to changes in climate, including variations in rainfall, ground and surface water levels, and glacial melt. Therefore, back-up may be needed in low-water years.

## Nuclear

China's first nuclear reactor went online in 1991. In January 2006, 9 reactors were in operation, with 30 more in the works. Nuclear plants generate about 2 percent of China's electricity. The Chinese government, planning an aggressive expansion of generated electricity, is conducting extensive research on nuclear-generated electricity. China

China's hydro-electric program is larger than the programs of Brazil, India, and Iran combined.

has reserves of almost 80,000 tons of uranium. For comparison, the United States has just over 300,000 tons, Russia has about 160,000 tons, and Australia, with the largest reserves in the world, has over 600,000 tons. Stated reserves vary greatly according to different sources.

## Chinese Electricity

According to Kahn and Landler in the *New York Times*, China is adding huge amounts of electrical power each year.

> China added 90 gigawatts of generation capacity in [2007], the third year in a row in which it will increase its power output by more than the total capacity of Britain. About 85 percent of those new power plants burn coal.

> The International Energy Agency, which had predicted as recently as a few years ago that China's carbon emissions would not reach those of the United States until 2020, now thinks China took the lead this year [2007].

Look out, world. That perfect storm cloud is as black as coal dust.


## INDIA

India—another huge country on the move, growing at over double the population growth rate of China—will likely surpass China in population in the next 30 years. India's population, 3.5 times greater than the U.S. population, lives in an area about one-third the size of the United States. This has disaster written all over it, although the Indian people are rapidly moving up the technology curve. Like the Chinese, Indians have the same growing appetite for energy for the same reasons:

- Rapidly growing population—about 1.1 billion in 2008 will become about 1.7 billion in 30 years (2038).
- Better life.
- Aggressively growing economy—from 1981–2001 the Indian economy grew a hefty 5.6 percent per year.

### Oil

India's proven oil reserves are 5.7 billion barrels, and the present rate of consumption is 1 billion barrels per year. Consequently, if India were to rely only on its own reserves, they would last less than 6 years at present use rates. India presently imports 70 percent of its oil. According to *Frontline*, a national magazine in India, India's oil demand is expected to grow from approximately 1 billion barrels per year in 2008 to 5 billion barrels per year in 2038.

Like China, India plans to get additional oil from producers around the world. Again, good luck. Consider the following article from Praful Bidwai appearing in India's *Frontline*. I quote this here because all countries could end up following the same path. If we do not solve our collective pollution/energy problem together, then we will all suffer together. We can see our own potentially frightening future in the following:

> High oil prices have confronted the Minister [Mani Shankar Aiyar] with a series of tough choices: How best to secure supplies of crude [oil] and gas even as their consumption rises by leaps and bounds? How to maximize India's leverage in the highly competitive global market for petroleum and gas? How to achieve energy security as the domestic production of oil falters? How to accelerate hydrocarbon exploration and improve recovery from existing Indian fields? How to protect the bottom lines of India's public sector superstars or Navratnas, the oil companies, while guarding the ordinary consumer and containing the inflationary impact of high crude prices? How to promote substitution of fossil fuels with bio-fuels like ethanol? In the larger context, how to promote the worthy cause—indeed, the environmental imperative—of energy conservation, while securing reliable energy supplies?...

> Given the political constraints under which such decisions are made, some slippages and suboptimal outcomes become almost inevitable....

> India's oil consumption, now about 2.25 million barrels/day, is estimated to rise, at present rates of expansion, to a huge 5 million barrels in five to seven years. This

**Other countries are much more interested in getting oil for themselves than in letting the United States gobble it up at what they think is an unfair, immoral, and greedy rate.**

should make all environmentally conscious citizens sick with anxiety. There is no way that India can or should sustain such high levels of energy consumption without causing enormous and irreversible damage to the global environment.

Such elitist consumption mania should have no place in a halfway sane society. True, our energy consumption per capita is under one-sixth that of the U.S. [JS: closer to one-thirtieth]. But that is no argument for emulating the U.S. Rather, it constitutes a case for reducing energy consumption in the U.S.

Yet, India's oil quest has been aggressive, even brash. Mani Shankar Aiyar has been holding road shows in numerous countries and signing up contracts for equity investment outside the traditional sources of West Asia. Our oil companies are looking to Russia, Latin America, and African countries from Angola to Chad, Niger, Ghana and Congo, to Sudan. Other targets include Ecuador, Sri Lanka, Iraq, and Venezuela. There is Myanmar of course; and above all, Iran with which a deal for a 2,600 kilometer (km) [1600 mile] gas pipeline through Pakistan is likely to be signed. India has signed a $2 billion contract for 20 percent holding in Russia's Sakhalin-I field. It wants to secure one million barrels/day from Russia alone.

India has emerged as China's main rival in grabbing oil contracts in as many countries as possible—following a long trail of rising powers, including imperialist states. For many decades until the 1960s, countries such as the U.S., France, and Japan used all kinds of methods to control oilfields and secure supplies. Britain divided up chunks of West Asia and created Kuwait to this end....

It is imperative that we take corrective action now—by saving energy, by discouraging private transport, and by pricing petroleum right to discourage its profligate consumption. Above all, we must promote non-polluting, non-greenhouse gas-emitting renewable sources like wind power. Wind electricity generation in India has now come of age and has become economically competitive...

We seriously need to promote renewable energy—as well as public or mass transport—through stiff, well-targeted levies on petroleum products and energy-intensive luxury goods such as cars and air conditioners. The alterative is unmitigated disaster.

The article above sounds desperate, as it should. Take heed, citizens of the world. The article describes the unfolding, accelerating energy problem, greatly exacerbated by population growth, facing all of us. The rest of the world is increasingly upset over U.S. excesses.

In India, 82 percent of its electricity is produced from fossil fuels, 14.5 percent from hydropower, and 3.5 percent from nuclear energy. This does not include a lot of energy produced at the point of use, such as solar energy used for hot water and cooking. India also plans an aggressive expansion of its wind-energy industry. Indians may have to slash their economic growth, because studies suggest that each 5 percent increase in the price of oil will cut India's growth rate by one-quarter of one percent. Also, no one has figured out what to do about India's out-of-control population growth.

India's population, 3.5 times greater than the U.S. population, lives in an area about one-third the size of the United States. This has disaster written all over it.

Consider the implications for India, as detailed by Ashish Vachhani in the *Hindu Business Line.*

In a world of growing petro-rivalry between nation-states, India has to catch up and outpace other major players in the global energy game. With the future of West Asia as a reliable crude supplier becoming uncertain since the US-sponsored "Operation Iraqi Liberation" (OIL), India is stepping up energy diplomacy with the countries in the South Asian region, Central Asia, Russia, Africa, and Latin America.

Washington has expressed its displeasure at New Delhi's newfound bonhomie with countries that the US has declared *non grata* in the context of our pipeline diplomacy with Iran. Though energy engagement with such states could be a risky strategy in terms of safety of our investments, *it appears to be a more secure arrangement than no energy security at all.*

Such developments make it absolutely necessary for India to push for restructuring the US-dominated international energy web and work towards creating a pan-Asian oil and gas grid to foster greater Asian energy security.

This article lays it on the line. When it comes to oil, the life blood of any nation's way of life, other nations really don't care at all what the United States thinks or does. Other countries are much more interested in getting oil for themselves than in letting the United States gobble it up at what they think is an unfair, immoral, and greedy rate. Energy concerns will increasingly drive and dominate nations' foreign policies and international activities. The following editorial, appearing in the *New York Times* on February 19, 2006, describes India's desperate problem.

Exploding at the seams with building, investment, and trade, India can hardly keep up with itself. City streets originally built for two lanes of traffic are teeming with four and sometimes five lanes of cars, auto rickshaws, mopeds, buses and trucks. This energy-guzzling congestion will only become worse as India continues producing fairly high-quality goods and services at lower and lower prices—from automobiles that cost only $2500 to low-budget airline flights for $50. All the while the population is growing and out of control.

India's president, A.P.J. Abdul Kalam, sounded exactly like President George W. Bush when he told the Asiatic Society in Manila that energy independence must be India's highest priority. "We must be determined to achieve this within the next 25 years, that is, by the year 2030," he said.

Sounds like "whistling past the graveyard" to me. Given India's runaway population problem, there is no way India will succeed. Either Kalam knows it, but is saying otherwise. Or Kalam doesn't know it. Either way, the enormity of the population problem defies simple solutions. Unfortunately, Kalam, like Bush, is far better at talking than at any real action to reduce energy consumption or to solve the real problem of accelerating depletion. When Bush made a recent long-planned trip to India, he visited a country that, like China, has begun to gear its international strategy to its energy needs. That is one of the biggest diplomatic challenges facing the United States, and right now the American strategy is inept because the real problem has not yet been defined. And we do not have a grip on our own energy problems.

The seeds of conflict have been planted. The United States must get on with solutions now and help others, as it is clearly in the country's best interests and in the best interests of avoiding diplomatic and armed conflicts over resources.

### Natural Gas

India's current natural gas reserves are 30 trillion cubic feet, and the present consumption rate is 1 trillion cubic feet annually. That gives India about 30 years to bone dry—assuming zero population growth and no economic growth.

## Coal

India, like China and the United States, has a lot of coal. India's coal, however, is generally low-grade. Proven reserves are about 100 billion tons, and India's consumption is approximately 500 million tons per year. With no population growth, no economic growth, and no imports, India will consume all of its domestic reserves in about 200 years. India has all types of coal, from lignite to bituminous, but most have high ash content and a low heating value. Coal is the major source of energy for generating electricity in India, where about 75 percent of the electricity is generated by coal-fired power stations. Also, other major industries in India, such as producers of cement, fertilizer, chemicals, and paper, rely on coal.

## Hydropower

India's hydropower potential is about half of China's, but still among the largest in the world. By the end of 2000, India's hydropower output accounted for approximately 18 percent of the electricity generated in India's public sector.

## Nuclear

By 1991 India had 11 operating reactors providing almost 3 percent of India's electrical energy. India, like China, plans rapid expansion of nuclear energy, including the development and use of fast neutron reactors. It is unclear if India will also use the recycling process called URanium EXtraction plus (UREX+, pronounced "UREX plus"). Plant operators will use India's own radioactive thorium for fuel because the nation has huge indigenous supplies. India's uranium reserves are about 60,000 tons. By 2020 India plans to have nuclear energy generate about 30 percent of its present electricity consumption, but it won't keep up with demand arising from predicted population growth. It seems certain that India's situation will become hopeless unless India quickly and decisively curbs its population growth.

India desperately wants President George W. Bush to wring approval from Congress for a misbegotten pact in which America would help meet India's energy requirements through civilian nuclear cooperation. (Note: this is a perfect example of how the Global Nuclear Energy Partnership could work. I discuss the GNEP at length in Chapter Ten.) With its eye on the nuclear deal, India recently bowed to American pressure and cast its vote at the International Atomic Energy Agency to refer Iran's suspected nuclear program to the United Nations Security Council.

## Solar

India has a lot of solar potential and has a large installed base of passive thermal solar systems for cooking and for heating water and air. The capacity of these systems is truly impressive, accounting for over 6 million square feet of collector area. India plans an additional 300 million square feet. India is also actively involved in research on solar photovoltaics to generate electricity, but considering the enormity of the need, the potential contribution from photovoltaics, while helpful, is minimal. Keep in mind that solar energy used for electricity must have a back-up power source.

## Wind

India's wind resources are quite significant. The country has on-land potential to generate 50–70 percent of its present electricity needs and 30–40 percent of its anticipated need in 2038. India ranks fifth in the world in wind generation. However, it is doubtful that even half of this potential can be realized because of present and (likely) future incapacity of the nation's electrical grid to hold and carry additional energy, not to mention the large land areas required. Also, recall that wind energy used to produce electricity must be backed-up by another power source when the wind is not blowing.

## Other Sources

Energy generated from tides, waves, and geothermal sources will not make a significant contribution.

## A Final Thought

This is a lot of information about India, but India's extreme energy issues foreshadow problems for all of us. India is a canary in a global environmental coal mine, a harbinger of things to come. India's situation teaches us a lot, because its problems have already arrived. As we watch, India leans into the first strong winds of the perfect storm.

> India's extreme energy issues foreshadow problems for all of us.

## THE WORLD—MISERY HAS a LOT of COMPANY

The overall situation in the world offers cold comfort, because there is plenty of misery to go around. *Assuming no population growth, no improvement in standards of living, no economic growth, no additional major discoveries of fossil-fuel sources, and no change in the consumption rates,* here is the situation the world faces:

- Oil will last less than 40 years (until, 2048).
- Natural gas will last about 65 years (until 2073).
- Coal will last 165 years (until 2173).

The global population will likely grow by about 45 percent in the next 30 years. Economic growth in many countries will be considerable. The economies of China and India will likely grow by 3 percent or more per year, while the United States, Europe, and others will grow at about 1.5 percent per year over the next 30 years. Most projections are for greater growth. I am certain growth will be seriously compromised by energy shortages even if the world makes a maximum effort to correct the shortages now.

*Considering the almost certain population growth and assuming only very modest global economic growth,* the world can expect the following:

- Oil will last less than 30 years.
- Oil will last less than 50 years if we double reserves by

tapping 1 trillion barrels of oil from unconventional sources and population continues to grow.

▶ Natural gas will last less than 50 years.

▶ Coal will last less than 75 years.

The above numbers take us to bone dry, all gone.

We will have big problems long before we completely run out. Consequently, I believe the world will see more gasoline rationing in the near future—and I believe the sooner the better. I recommend it immediately. There is already de facto rationing going on in the world. India, for example, has a quota limiting (rationing) the amount of some fuels that people can buy at deeply discounted prices. Rationing is on its way in part because some governments are already desperately trying to keep prices reasonable to consumers while world prices increase. According to *Economist* magazine, only one-third of the 48 developing countries let the markets set fuel prices. The governments of Yemen and Indonesia, for example, spend more money holding down the price of fuel through subsidies than they spent on health and education. This again has a ring of desperation, and it will only get worse. The first gusts of the perfect global storm are blowing.

I'll bet you think I've lost my mind. These numbers can't possibly be accurate, right? I thought so, too. The numbers are so bad that they are shocking at first and then unbelievable. But believe them. The real shock for you and me should be the fact that the United States and the world have done so little about this *impending, on-going* disaster. Energy independence? When and how long will it take the United States to get there?

> Oil and energy are intimately and inextricably tied to war, jobs, and economic conditions.

During the 2006 U.S. midterm elections, energy was essentially a non-issue. Gay marriage, war, taxes, abortion, fuel prices, immigration, and ball parks were talked about with the assumption that these issues were more important to our politicians and to us. Not much has changed as the 2008 presidential campaigns unfold. Iraq, jobs, and the economy dominate. Don't people see how oil and energy are intimately and inextricably related to all 3 issues?. I suspect other world leaders have similar reluctance to deal with this issue. The economy, war in Iraq, and the price of fuels are related to energy, but they don't go to the core of the energy issue. None of these issues will matter much when we get a little closer to running out of energy, and others refuse to sell it to us or will only sell it at prices that will destroy our economy.

You can of course disagree with the above numbers based on differing projections (there are many), but the overall picture won't change much. A difference of 10 or even 20 years in the overall picture

> **Keep in mind that any proposed solution that is without quantification and a firm timetable is not a plan; it is only a dream.**

is really meaningless. The United States and the world are running out—and soon. Surely your grandkids will be left holding the bag— an empty bag, that is.

The recovery of oil from oil shale and other unconventional sources could make us more energy independent and could buy us some time. Yet even if we double the present worldwide reserves from 1.1 trillion barrels to 2.2 trillion barrels, we could still run out if nuclear and other renewable energy sources, as well as transportation reform, are not aggressively pursued in the next 30 years. (Note: Included in the 1.1 trillion barrels of conventional oil is 100 billion barrels of conventional oil that is yet to be discovered.) The arithmetic is really quite simple. Keep in mind that any proposed solution that is without quantification and a firm timetable is not a plan; it is only a dream. The time for more studies and glib solutions is over. This book proposes a very affordable solution *and* a timetable.

America will be losing its position as the most productive economy on the planet to China within a decade or two and to India probably some time in this century. No longer will the United States be as admired or feared as in the past. For the first time in history, according to the 2005 A.T. Kearney *Foreign Direct Investment Confidence*

*Index, the United States ranked as only the third best country for investment, behind China and India.* This ranking held steady in the 2006 and 2007 Indexes too. I would be very happy for all 3 countries to be energy independent for obvious reasons.

Also remember that the United States is twice as far from the Middle East as China, 3 times farther than Europe, and 4 times farther than India. World geography offers no comforts if push comes to shove.

## BOTTOM LINE

- ◖ There is no place for the United States to turn to for energy relief, except to look to our own resources and resolve. The United States must turn to available renewables.

- ◖ All countries should help each other. The good health of our common home must be the common goal. The United States and other wealthy nations must help other, less fortunate countries become energy independent for mutual benefit.

- ◖ Above all, we must avoid the idea of "every country for itself and to heck with all others." The world must avoid war, the ultimate, incredibly costly, zero-sum game.

- ◖ We must encourage and help all countries to develop a 30-year plan to energy independence. After 30 years, the cost of transition goes up on a very steep curve.

*Part Two*

**SOLUTIONS**

# Chapter 6

## SOLAR ENERGY

The sun provides limitless energy. The sun radiates over the face of the earth thousands of times more solar energy than humanity needs. Indeed, all energy—including the energy from oil, gas, and coal—comes from the sun through photosynthesis, the process whereby the sun (plus carbon dioxide and water) greens the plants that eventually die. As the sun fuels the growth of all plants, it also creates temperature differences that generate wind, cause evaporation, and drive the water cycle. In this way, solar power comes to us in many forms—fossil fuels, direct sunlight, wind, hydropower, and biomass. I discuss these sources elsewhere in the book, but this chapter addresses only direct sunlight as an energy source.

### DON'T BELIEVE the HYPE

Don't believe the wild projections about how direct solar energy will satisfy a large portion of U.S. electrical energy needs in the next 50 years. Both the use of direct solar energy and wind energy are experiencing very rapid growth, yet their contribution to the production of electrical energy remains miniscule. Some people deeply believe that all energy problems can be solved by direct solar energy and wind power. Too bad these people lack supporting numbers to back up their hopes. Do you recall any of the grand proclamations that predicted a large portion of U.S. energy—up to 50 percent—would come from renewable sources by 2005? We keep hearing such predictions, yet utilities still propose and build coal plants. Unsubstantiated hype only confuses the public, so don't believe it until the advocates present some solid evidence: Who? When? Where? How much?

## Pie in the Sky Plans?

I strongly advocate solar-powered electrical generating systems in two specific applications. First, direct solar energy can produce heat and provide peak power in response to peak demand (peakload), regardless of what kind of power provides the baseload. Every kilowatt hour generated by solar power reduces fossil-fuel pollution. Second, direct solar energy can often be the best choice for producing off-grid electricity at the point of use.

The solar-energy industry has broader plans. The industry drafted an ambitious plan requiring a gigantic effort to supply half of the *new* electrical energy required by the United States between now and 2025. This much electrical energy will account for about 10 percent of the total U.S. consumption that year. This significant amount is more than all the electrical energy consumed in Italy and 1.5 times all the electrical energy consumed in Mexico. I respect and encourage the plan, but I am skeptical.

How might direct solar power generate 10 percent of U.S. electrical energy needs? Getting to 10 percent will be a daunting task. It's doubtful that the United States will get there by 2025. However, the public should not allow this goal to fail through a lack of funding. Besides, if the United States can't generate 10 percent of its electrical needs from solar energy by 2025, then how will the country ever get to 20 or 30 percent—much less 100 percent—in time to avert the impending fossil-fuel-related economic meltdown?

I believe that over a longer time period, say 100 years, direct solar energy will grow in step with advancing solar technology. By 2100 direct solar energy may begin to become the world's major source of electrical energy. A strong research effort to this end should continue. Costs must be reduced, and construction materials must be found that are more abundant and more environmentally friendly. For example, some high-efficiency photovoltaic cells require exotic materials, and the quantities of such materials are often insufficient to support a massive deployment of solar power.

Photovoltaic solar installations are quite expensive, but they are competitive in some applications, particularly where the generated electricity is used on site, thus by-passing the grid. Per kilowatt, the installed cost of photovoltaic panels is 3–4 times the installed cost of windmills and about 7–8 times the installed cost of nuclear energy.

In short, solar will play a minor but important role, but for many reasons direct solar energy (and wind energy) will probably never make a large dent in the world's massive energy requirements or provide an answer to U.S. energy needs. Beware of predictions to the contrary.

When doing research, you soon learn that sources often offer circular arguments that quote each other and quote the media sources who quoted them in the first place.

## Back-up Energy

Solar power does not reduce a utility's baseload power needs, because there will be times (as when the sun is not shining) when 100 percent of the needed energy must come off the grid. One conclusion is clear: Direct solar energy (or wind energy) always needs a back-up source of energy for when the sun does not shine (or the wind does not blow). The remedy would be a super battery or some other efficient energy-storage system, but they aren't here yet. In the meantime, back-up energy must come from a fossil-fuel power plant, a hydro-electric power plant, or a nuclear power plant. Of course, if the back-up is nuclear, then there is no need for solar or wind energy in the first place, since nuclear energy is more economical and cleaner.

## Background Reading

I recommend two well-written books every solar enthusiast should read. Mr. Travis Bradford's book *Solar Revolution* (MIT Press, 2006) strongly advocates for solar energy and makes a good case for distributed solar power.

Dr. Howard C. Hayden's book *The Solar Fraud: Why Solar Energy Won't Run the World* clearly conveys his criticisms. If you are an avid proponent of solar power, then Dr. Hayden will likely enlighten and perhaps surprise you. Bradford and Hayden come to their conclusions using different assumptions, but Dr. Hayden's position is more founded in mathematical analysis.

Mr. Bradford develops his position from a number of assumptions that are yet to be proven, and his objections to nuclear energy are way out of date. Also, it is not clear what he recommends for baseload power. In one instance he shows wind as the baseload provider, but this will never happen since wind is itself an intermittent energy source. Bradford also claims that "developments in the photovoltaic industry over the last 10 years have made direct electricity generation from photovoltaic cells a cost-effective and feasible energy solution." I don't believe that even the most optimistic solar scientist would agree with this claim *unless* Bradford added "in some applications." *If the cost of pollution is included* in the calculations, then there is no doubt that solar energy in all of its forms is cost competitive with fossil fuels' full life-cycle cost—that is, the cost of fossil fuels from cradle to grave, including environmental damage.

Dr. Hayden rebuts much of what Mr. Bradford claims. Not only does Hayden discuss solar energy in all its forms, but his comments also clear up much of the misinformation that finds its way into the popular press about energy. Hayden also clears up other wild statements released by some non-governmental organizations—I call them "press release factories"—often quoted in the press.

## How Much Do You Pay For Electricity?

Dr. Hayden also illustrates the distortion in energy preferences and prices caused by subsidies. While in some cases subsidies are necessary, policymakers should be ever-vigilant of the unintended consequences that often arise when subsidies are based on political considerations, are too complex, or have no basis in common sense.

But what does that mean? Do most consumers know how much their electricity costs? Polls show that most consumers are willing to pay extra for clean electricity. If most U.S. consumers who now pay 7¢–10¢ per kilowatt hour had to pay a bit more for clean, renewable energy, then they would still be paying a lot less than the cost of continuing the use of fossil fuels and much less than the price paid for electricity in other countries (see Figure 6.1).

## Figure 6.1. Electricity Costs in Various Countries

| Country | U.S. Cents Per Kilowatt Hour, 1997 | U.S. Cents Per Kilowatt Hour, 2004 |
|---|---|---|
| Brazil | n.a. | 9.3 |
| China | 4.6 | n.a. |
| France | 13.4 | 14.1 |
| Germany | 15.9 | 18.7 |
| India | 3.2 | 4.5 |
| Iceland | 13.1 | 17.3 |
| Italy | 16.0 | 19.1 |
| Japan | 20.7 | 19.6 |
| Netherlands | 13.0 | 22.1 |
| United Kingdom | 12.5 | 13.8 |
| United States | 8.4 | 9.0 |

Source: U.S. Department of Energy, U.S. Energy Information Administration (February 28, 2006).

It becomes obvious that the more electricity costs, the greater are the incentives to choose solar energy. Japan and Germany are leading the way for solar. While Europe, Japan, and much of the rest of the world are increasing their use of solar energy every year, U.S. use has actually declined.

## DIRECT SOLAR ENERGY

The countries making the most use of solar energy are Japan, Indonesia, Australia, Mexico, the United States, Northern Africa, and Western Europe. Sadly, the United States surrendered its leadership in this growing industry to Germany and Japan. In 1997 U.S. manufacturers supplied 40 percent of the world market, yet in 2003 U.S. producers contributed less than 15 percent—a dramatic drop in just 6 years. The U.S. share of the worldwide deployment of photovoltaic solar energy is a paltry 14 percent (see Figure 6.2). The industry generated $4.7 billion in revenue. Due to a lack of leadership and foresight, the United States is rapidly being shut out of this fast-growing industry and is surrendering technology and jobs to foreign countries.

Increasingly, policies in Europe and Japan are driving the development of solar technology and markets. However, an economic fact of life is that both Japan and Germany pay more than double what U.S. consumers pay for electricity. Japanese and Germans simply have more incentive to make solar energy and other renewable energy sources work—and that's okay if U.S. citizens and entrepreneurs are smart enough to hitch a ride on their development work.

Over 1000 megawatts of photovoltaic solar power are installed worldwide, enough capacity to supply the electrical needs of a city of 700,000 people when the sun is shining. This excludes industrial needs and the obvious need for back-up capacity. Put another way, this amount of energy is the equivalent to 500–600 typically large-size windmills, or 2 million barrels of oil, or one-half million tons of coal. While this may sound like a lot, it is almost nothing, less than one-tenth of 1 percent of total energy consumption in the United States.

The United States is rapidly being shut out of this fast-growing industry.

## Figure 6.2. Worldwide Photovoltaic Shipments, 1988–2004

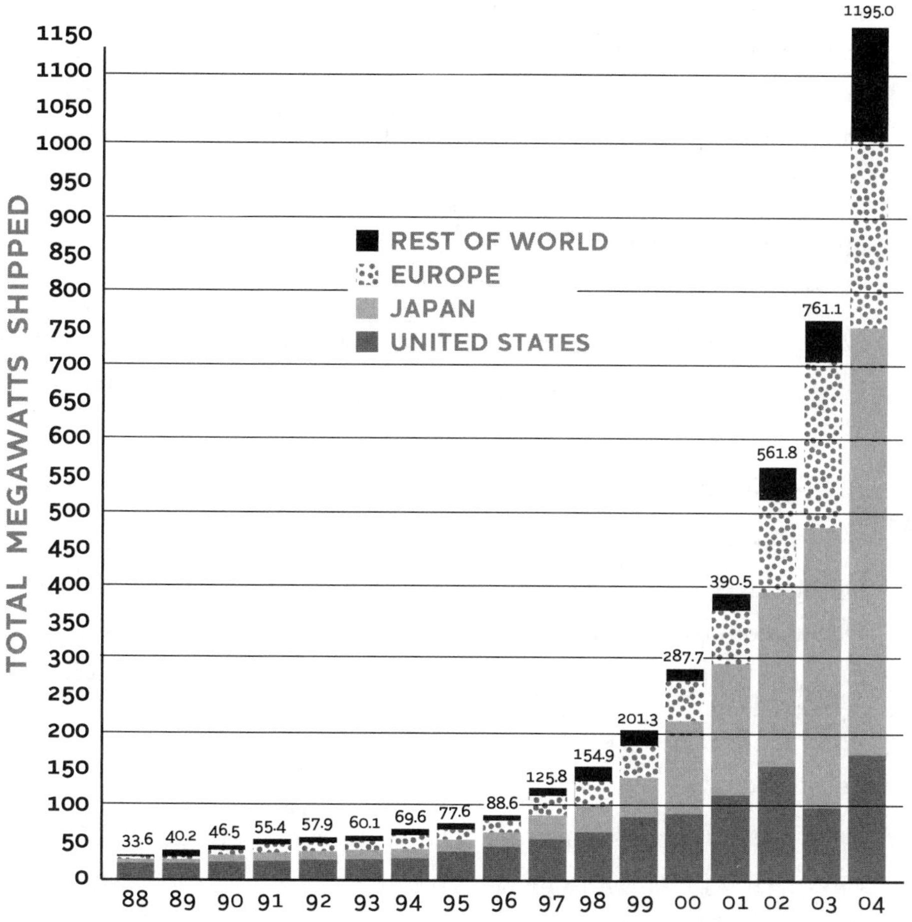

Electricity generated by photovoltaic units represents only 0.02 percent of world electricity consumption (2004).

Source: National Renewable Energy Laboratory, 2005.

## Worldwide Distribution

In the United States there is enough sun in all 50 states to beneficially deploy solar systems.

### Best areas

The best areas are between latitudes 15° and 35° north and south. These areas have few clouds, little rainfall, and enjoy over 3000 hours of sunshine per year. Approximately one-half of Japan, the most active country in solar technology, is in this area. In North and Central America, these latitudes include the area from the southern tip of Mexico to Kansas. Since many developing countries fall within this zone, the use of solar energy should be greatly expanded in those nations.

### Moderately favorable

The equatorial area from 15° north to 15° south latitude gets plenty of sunshine, but because of high humidity and greater cloud cover it is second best. The area receives about 2500 hours of sunshine per year.

### Less favorable

The third best area lies between latitude 35° north and south to latitudes 45° north and south. In the U.S. this includes an area north of Oklahoma to the middle of Minnesota. This also includes southern Europe, northern China, and South America between the middle of Bolivia to the middle of Argentina. Half of Japan is in this area, too.

### Least favorable

The least favorable area for solar is the area north of 45° N and south of 45° S. This area includes the greater part of North America including Canada, northern Europe, and Russia. This area has extensive cloud cover, and the energy radiating from the sun is more diffuse, yet these conditions do not preclude the use of solar energy, particularly if the cost of solar-energy systems drop as predicted in the next 30 years. Most of Germany, one of the most active solar countries, is in this area.

The interplay of subsidies, lobbying, and campaign contributions often skews legislation and makes it difficult for non-fossil-fuel industries to get traction.

## Solar Power in Key Markets

In the past decade firms in Japan installed more than 750 megawatts of grid-connected systems on homes and businesses. In Germany firms installed more than 400 megawatts. In the United States the solar-energy industry installed about 340 megawatts of off-grid and grid-connected systems. The additional 340 megawatts in the United States contributes far less than 0.1 percent to total U.S. electrical energy needs.

Many parks and other public places in California get a large portion of needed electricity from solar-powered systems. The U.S. federal government—unlike national governments in other major solar markets, such as Japan, Germany, and other European countries—provides no coordinated nationwide incentive for solar energy. For that matter, the U.S. federal government has shown no reasonable long-term commitment for any renewable energy. Oil, gas, and coal still reign.

The regulatory issues and tax incentives in the United States to develop new, non-polluting energy sources are convoluted, confusing, and often downright stupid. This counter-productive mess results from the influence special interest groups and powerful corporations can bring to bear on our political leaders. Firms contribute millions of dollars to the campaign funds of both parties, and the donors expect favorable legislation, or at least little unfavorable legislation. Such legislation, often quite beneficial to specific firms and industries, is equally often detrimental to the common good. In 2004, individuals and businesses involved in the fossil-fuel industry contributed over $30 million dollars to politicians and political parties. In 2006, these groups also spent approximately $40 million in their lobbying efforts. The tax breaks and research support lavished upon the fossil-fuel industry by the U.S. federal government amounts to billions of dollars annually. The

> The U.S. federal government has shown no reasonable long-term commitment for any renewable energy. Oil, gas, and coal still reign.

interplay of subsidies, lobbying, and campaign contributions often skews legislation and makes it difficult for non-fossil-fuel industries to get traction.

## Advantages of Solar Power

Solar power offers at least 8 clear advantages over fossil fuels.

▷ **Predictable Electricity Costs.** The fuel for solar systems—light from the sun—is free. The cost will never rise.

▷ **Lower Cost for Peakload Power.** Solar power peaks when demand for electricity peaks—that is, on hot summer afternoons when utilities have trouble meeting demand for air conditioning. During summer peakload periods, when transmission is constrained and utilities are generating electricity in the most expensive fashion, solar energy can substantially increase operating margins and decrease costs.

▷ **Reliable Power.** Solar-power systems, particularly when integrated with reliable energy-storage systems, can provide electricity during even the worst power outages. Energy-storage batteries and other electricity-storage systems should become a major focus of U.S. research and development. In the future, thousands of solar-power systems distributed throughout the electrical grid could reduce vulnerability to equipment failures. At a critical moment, thousands of small solar systems are less likely to fail than a single, large power plant, transmission line, or transformer.

▷ **Abundant, Secure Domestic Energy.** Sunshine is plentiful and secure. Solar collectors on a 150 x 100-mile area in the Southwest region of the United States could generate all the electricity U.S. customers need. By 2040 a 170 x 100 mile area would be required. Alternatively,

Travis Bradford claims in *Solar Revolution* that solar-power systems on roofs, parking lots, and other developed land across the nation could generate all the electricity the United States needs—now, in 2030, and in 2050—without building on the nation's open spaces. I don't believe this will ever happen because of the many practical and political limitations.

❯ **Less Air Pollution.** Solar power produces no on-site air pollution. Every residential photovoltaic system reduces greenhouse gas emissions as much as removing one-third of a car from the road.

❯ **Lower Water Consumption.** Solar power uses 98 percent less water per megawatt hour generated than the most efficient electricity-generating plant burning natural gas.

❯ **Easy to Site.** Solar power is one of the few generating options to work well almost anywhere—in crowded urban centers, suburban housing developments, or rural farmhouses. On-site solar power also avoids expensive upgrades to substations or replacement of buried power lines. On-site solar power has no moving parts, requires no water, has no emissions, and can be smoothly integrated into a building structure. These characteristics are ideal for a power source that delivers electricity where it is most valuable—at the point of use. This is true of photovoltaic systems, but not true of large sun-concentrating systems.

## Challenges Accompanying Solar Power

❯ **Intermittent Sunlight.** This problem can be reduced by improvements in batteries and energy-storage systems.

❯ **Back-up Power.** Solar-power systems need 100 percent back-up when the sun is not shining or when the sun's energy is insufficient to keep up with demand. Solar power makes a lot of sense if the back-up plant is fired by fossil fuel, but solar power makes less sense if the back-up is a nuclear power plant, which is cleaner, renewable, and actually a lot less expensive.

▶ **Space Concerns.** Solar energy is diffuse and takes a lot of space. The most likely locations are deserts, where land values are low and the sun is most intense. However, deserts are far from major population centers, so a large-scale extension of the electrical grid would be very costly. In the United States the land required to supply all of the electricity required today from photovoltaic solar systems with today's technology is very large. This area is well in excess of 10 million acres, about half the size of New York State. By 2037 these numbers would be 45 percent higher. By 2050 they would be approximately 60 percent higher, if population continues to grow as predicted.

▶ **National Electrical Grid.** Extensive use of solar energy will likely require very costly extensions of and modifications of the existing U.S. national electrical grid. This could render solar energy cost-prohibitive.

▶ **Installation Costs.** Except for most passive systems for space and water heating, the cost per installed watt of solar power is much greater than wind and nuclear. (See the last chapter for specific comparisons.) This cost is coming down in step with market growth and technological improvements, but there is still a long way to go.

▶ **Maintenance.** Expect high maintenance costs for systems spread over large areas. In general, I think maintenance costs have been underestimated.

▶ **Durability.** Can solar-power systems, especially if spread over large areas, survive the elements, hail, ultraviolet light, lightning, and other extreme weather conditions?

## MORE LIGHT on SOLAR ENERGY

One can use solar energy by directly tapping the sun's heat for useful purposes, or one may convert the sun's energy into electricity by using photovoltaic systems or by using mirrors to concentrate the sun in Concentrating Solar Power (CSP) systems. Photovoltaic systems capture sunlight, which reacts with special materials to produce an electric current. CSP systems use mirrors to concentrate the sun's heat energy to boil water to produce steam to drive an electricity-producing turbine. To get needed energy most efficiently, one must match the need with the best system.

### Directly Tapping Solar Heat

*Active solar heating* systems absorb heat via solar collectors, which are generally located on a building's roof. Sunlight warms a liquid within the collectors. As the equipment circulates the liquid throughout the building, the liquid's heat warms the building or can be used simply to heat water.

Buildings with *passive solar heating* systems collect the sun's heat during the day, then release the heat when needed at night. To better absorb heat, some buildings are strategically designed with large windows in strategic locations to collect heat for winter heating, and awnings to block the sun's heat and light in the summer. My own home is passively heated by the sun. Passive solar systems used to heat water and living spaces have always been economical, particularly if these systems are designed into the structure of the building. Designers and architects must become more aware of energy issues and get better at including energy-saving systems into our homes, offices, and other buildings.

### Converting the Sun's Energy into Electricity
*Photovoltaic Systems*

Photovoltaic systems directly convert light into electricity using semiconductor materials. They have no moving parts, make no noise, need only sunlight for fuel, and produce no on-site pollution. Since photovoltaic systems operate during the day, when demand is highest, they are ideal for supplying peakload power on the hottest days, when air conditioning needs are the greatest, for example. Photovoltaic systems are typically modular, making them relatively easy to install and expand. Photovoltaic production is doubling approximately every 3 years. For every doubling, experts predict the cost of photovoltaic systems will fall by 10 percent.

When electricity is produced at the point of use, the method is called a *stand-alone* or *distributed generation system*—no electrical grid is required. Distributed generation is always a sensible solution for local or specific electrical needs, particularly where the grid is not accessible. A stand-alone system is especially efficient since it suffers no transmission losses. Photovoltaic systems can also be connected to the grid. A *grid-connected system* not only draws from the grid when the sun is not shining, but also sells excess energy to the grid. It is easy to switch the output from a photovoltaic system to the grid and vice versa.

In any case, to have electricity at night or at times when the sun is not shining, a back-up generating or storage system is required. The airport at Glen Canyon National Recreation Area, Utah, is powered mostly by photovoltaic systems. Similarly, some national parks use photovoltaic systems to generate clean non-polluting electricity to support most of their activities.

### Stand-alone photovoltaic systems.

Around the world stand-alone photovoltaics are providing electricity to many remote areas. Indonesia, a nation of islands, is using stand-alone photovoltaic systems so neither the government nor utilities confront the staggering expense of connecting the many islands to a central electrical generating plant and a distribution grid. In some countries, including India, customers can buy complete home photovoltaic systems, including wiring, appliances, and solar panels. Most photovoltaic systems can be mounted atop a roof, where they are hardly noticeable. In fact, solar shingles that can be mounted with regular shingles are available on the market.

While photovoltaic electricity contributes clean energy, it does not represent a significant contribution to the world's energy needs. This could change as photovoltaic panels get more efficient and less expensive, and as better, more robust energy-storage systems are developed.

### Grid-connected, electricity-distribution photovoltaic systems.

The often sun-drenched states of California, Arizona, Texas, and Colorado are making the most use of grid-connected photovoltaic systems. Some of this activity was stimulated by the U.S. government's million-dollar Solar Roofs program. Grid-connected photovoltaic systems, which come in sizes of 1-megawatt or larger, provide electricity to large buildings or multiple buildings. When solar panels are planned as part of a roof system, costs can become more favorable since part of the expense of the conventional roof is eliminated. Plants that generate at least 1 megawatt of solar electricity—about half the rating of a fairly large, standard windmill—are presently in use in at least 6 countries: Germany, India, Italy, Japan, Spain, and the United States.

> While photovoltaic electricity contributes clean energy, it does not represent a significant contribution to the world's energy needs.

### Photovoltaic panels and pollution.

Producing photovoltaic panels requires considerable energy. Surprisingly, it takes approximately 3 years of electricity generated from a photovoltaic system to break even with the electrical energy required to produce it. Then there is the problem of safely disposing of some nasty materials when the cells are no longer functional.

### Concentrating Solar Power (CSP) systems

These systems, generally large and grid-connected, consist of mirrored reflectors concentrating the sun's energy. This energy heats water to produce steam to drive turbine generators, as in a conventional plant fired by coal or natural gas. The big difference is that CSP plants are essentially non-polluting and can approach twice the efficiency of a photovoltaic system. There are 3 main designs of CSP collector-concentrator systems: a large dish, a parabolic trough, and a central receiving tower. Dish systems typically generate only 5–25 kilowatts, a small amount of energy, so I will not discuss them.

### Parabolic troughs.

The parabolic trough is the most advanced CSP system. Parabolic troughs, which can be quite long, concentrate the sun's heat at a long fluid-filled pipe at the focal point of the curved trough. The heated fluid in turn heats water to create steam. The trough automatically rotates east to west to stay lined up with the sun. The largest trough system is located in the Mojave Desert, where 9 individual trough systems feed about 350 megawatts of electricity to homes in southern California. At a 35-percent capacity factor, the electricity generated would partially service approximately 100,000 homes.

### Tower systems.

Central-receiving-tower systems concentrate the sun's energy at a central, fluid-filled tank mounted on a huge tower. Around the tower are a multitude of mirrored reflectors that move with the sun and focus its rays onto the boiler tank. The fluid, generally water, is heated to approximately 2500 degrees Fahrenheit, which in turn produces steam to drive a turbine. This system has the ability to store energy more effectively than the dish or the trough system for times when the sun doesn't shine. While not many of these systems have been built, there is considerable interest in this type of solar system. Tower systems are in the early stages of development compared to the more mature parabolic trough systems. I have high hopes for tower systems, particularly since they have energy-storage capability.

## SOLAR ENERGY in the FUTURE

Technological advances are driving down the costs of solar energy and improving basic efficiencies and economies of scale. Although costs will fall, I believe costs will remain a problem and could limit the deployment of major solar plants. Nevertheless, solar energy will play a role in the drive to U.S. energy independence by 2040.

There are many opportunities. One application uses recently developed flexible photovoltaic cells in the fabrication of tents deployed in Iraq to provide energy for U.S. troops. Also, every major building in the United States and around the world should be studied for solar opportunities. One very interesting and outstanding application of photovoltaic energy is a skyscraper at 4 Times Square, which incorporates photovoltaic cells in the south wall. The cells act as exterior glazing material and as an electrical source for the building. The panels extend from floors 37–43. This dual use makes this installation one of the most cost-effective urban solar arrays ever installed. Cities with high electricity costs provide excellent opportunities for clean, secure, and affordable solar electricity.

Engineers will learn more about solar energy as they install more systems and stimulate the entrepreneurial and innovative spirit of our business and scientific leaders. Patent activity involving solar energy suggests such stimulation is already happening. The greatest need, however, is more political and public enlightenment on energy matters.

## THE SOLAR ROADMAP, SEPTEMBER 2004

The Solar Energy Industries Association beautifully presents a possible future for solar energy in a booklet entitled "Our Solar Power Future." The booklet collects the wisdom and experiences of a consortium of utilities, universities, public policy leaders, energy expert consultants, photovoltaic and inverter manufacturers, system integrators, several of our National Laboratories, and a host of other stakeholders. The "best of the best" put the plan together and dedicated their joint resources to implement the plan through investments in technology and market development. The booklet outlines a plan for deploying photovoltaic systems in the United States through 2025 and beyond. The goal, as I state above, is for solar power to provide half of all *new* U.S. electricity by 2025. The plan also seeks to deliver more jobs, a cleaner environment, and more secure domestic energy. Yet this bright future will not happen without solid investment and public backing.

## Investing in the Plan

Investment decisions over the next decade for research, new manufacturing, and creating new markets will determine where solar power will thrive and where it will merely survive. I believe U.S. leaders should do everything possible to accelerate the plan's timetable by doubling the funding for research and development and by funding the recommended electrical energy surcharge of 2.5¢ per kilowatt hour with a predictable, uncomplicated incentive program. Effective policies sustained over time will increase the production of solar power, dramatically expand markets, improve technology, and reduce costs. Programs in Germany, Japan, and California prove it.

The solar industry's roadmap calls for eventual capital costs of $2300 per installed kilowatt—a tough goal. Today the capital costs of solar-power systems are $6000–7000 per kilowatt, and going down. Even at a capital cost of $6500 per kilowatt, solar power is a better buy than a coal-fired plant, when environmental destruction costs are included.

It should be obvious to all that investments in solar energy, along with wind power and nuclear power, would be the best and wisest energy investments the country could make. This book, especially the last chapter, proves that point beyond a doubt. Beyond the virtues of a cleaner and healthier environment, helping *all nations* become energy independent is surely the single most important thing humans could do for world peace. Also, it is clear that with clean, inexpensive energy, the lives of most people on earth will be improved, poverty diminished, and education made more widely available in parts of the world that have known only despair. To achieve a more peaceful world, the affluent nations must help less fortunate countries.

## The Plan to Reach Tomorrow

The solar industry's plan began in January 2005. The worldwide industry is growing more than 30 percent annually, but it is increasingly dominated by German and Japanese companies.

However, with robust investments in research and market development, the picture changes dramatically. Expanded support for research and development offers considerable opportunities to reduce costs quickly. Decreased costs open up new market opportunities, which, in turn, expand solar power shipments and help to further reduce costs. Most of the current cost is depreciation of equipment and interest on the investment. However, all indications suggest that solar power is still in the early stages of exploiting its potential.

Scientists worldwide agree that solar-power technologies can become significantly more efficient, more reliable, more durable, and less expensive. In technology areas such as plastic solar cells, nanostructured materials, and dye-sensitized solar cells, scientists see the potential to leapfrog far beyond current crystalline silicon and thin films and to dramatically lower costs and raise performance. The European Renewable Energy Council projects that by 2040, solar power *could be* the largest source of renewable energy generating electricity, supplying over one-quarter of worldwide electricity consumption. This projection is extremely aggressive. I believe the solar-energy industry will be frustrated in this effort given the time, cost, and energy back-up systems required. In fact, I believe given the urgency, this goal

> It is clear that with clean, inexpensive energy, the lives of most people on earth will be improved, poverty diminished, and education made more widely available in parts of the world that have known only despair.

is totally unrealistic.

To make the United States competitive with European and Japanese rivals, the solar industry's roadmap calls for the United States to proceed by gradually increasing its annual investment toward research and development on solar power to $250 million by 2010. I believe this investment should be doubled to $500 million, since technological innovation will be the only way the United States can become more competitive. Solar-power research has helped reduce solar-power costs by nearly 50 percent in a decade, and such research is essential to making solar power more broadly competitive in the next decade. Coupled with this effort should be an aggressively funded program to develop better batteries and energy-storage systems.

## BOTTOM LINE

▶ Per the solar industry's plan, do whatever it takes to get to 10 percent solar-generated electrical energy in the United States by 2025. This must not suffer for lack of funding.

▶ At the same time immediately increase funding for solar energy research to at least $500 million annually.

## Chapter 7

# WIND

The economics of wind power are reasonably favorable, and some very smart Europeans are showing the world how to deliver a big percentage of wind power to a relatively large electrical grid. The amount of wind power installed over the past decade has grown by roughly 25 percent per year. In 2006 wind supplied roughly 1 percent of the total electricity generated worldwide, and 75 percent of this capacity is installed in Europe, where electricity costs are high. In contrast, wind generated less than 1 percent of the electricity consumed in the United States in 2006.

Is deliverance at hand? Will wind power save the day? There is no doubt that wind power is more economical than fossil fuels, if all social costs are included. Yet the inherent limitations of wind power restrict it to a relatively minor role in generating the electrical needs of the United States—about the same prospects as solar energy in the short term.

I don't believe that wind power will ever account for more than 10 percent of total energy production in the United States. I also believe that arriving at the 10-percent-production mark will be an almost impossible task. Of course, the wind-power industry says we can do a lot better, but, as with solar power, let's see the 10 percent first. I do, however, support plans to generate 10 percent as soon as possible. After that, the United States will be in a better position to determine the long-term future of wind.

Neither wind alone nor sun alone—nor a combination of both—will solve the energy problems confronting the world and the United States. Yet wind will be *part* of the solution.

## WIND BASICS

Wind energy, like direct solar energy, is eternal and diffuse. Indeed, wind is a form of solar energy. Wind blows because the heat of the sun creates temperature differentials that cause the air to move. People have used wind for energy for at least 3000 years, mainly for grinding grain, pumping water, and driving sailing ships. Windmills have been fairly common since the thirteenth century, but fell into disuse with the advent of cheap fossil fuels. Wind turbines to generate electricity didn't come on the scene until the late nineteenth century in the United States and Denmark. They became almost "extinct" when electrical grids were extended to rural farms and communities. Now, however, the high price of fossil fuels and environmental concerns are driving the development and deployment of modern wind turbines. We know tornados, hurricanes, and straight-line winds cause billions of dollars of damage each year, but now we can get a little back as we harness the wind to drive windmills to produce electrical energy.

### An Introduction to Wind Turbines

The elegant 1.5–2.0 megawatt windmills are engineering marvels. One megawatt is enough electricity to power about 250–300 homes, though the actual number varies with the size, design, and location of the homes. One very popular size is a 1.65-megawatt unit. Such a windmill in a reasonably windy area can produce enough electricity for approximately 450 average U.S. homes. Of course, these homes must get energy from the grid when the wind is not blowing. The 1.65-megawatt windmills stand about 330 feet high and support blades having a diameter of about 260 feet—taller than a football field is long, with the diameter (span) of the blades almost as long as a football field. When you get close, they are truly majestic and awe-inspiring structures.

Why are the blades so long? Longer blades generate more power. A windmill's energy output varies as the square of the diameter of the blades. If you double the length of the blades—all other parameters being equal—you get 4 times the energy output, and if you triple the length you generate 9 times more energy.

> Neither wind alone nor sun alone—nor a combination of the two—will solve the energy problems confronting the world and the United States. Yet wind will be part of the solution.

The seemingly simple design of a windmill is actually quite complex. The preferred design attaches 3 simple blades to a shaft that drives a turbine. This design incorporates many engineering trade-offs and compromises, and how these trade-offs are managed determines the windmill's overall output and efficiency. To illustrate, let's consider the standard 1.65-megawatt unit spinning at approximately 15 revolutions per minute (rpm) and geared through a transmission to drive the turbine. Fifteen rpm seems quite slow, but because the blades are so long, the speed at the tip of the blades is approximately 140 miles per hour. If you double the diameter, then you double the tip speed. Arriving at the optimum design is an ongoing challenge.

## Capturing Wind Energy

No matter how passionate you are about wind energy, you cannot change the laws of physics that limit the amount of usable wind power to be extracted.

Windmills extract only a small portion of the energy from the wind that hits the turbine blades. For example, if the power density facing the wind is 300 watts per square meter of land, then you can extract from this wind only 4 watts per square meter of land. According to Howard C. Hayden (*The Solar Fraud*, p. 144), the power density divided by 75 determines the extractable power [300/75 = 4].

Since wind turbines cannot be lined up directly behind each other, rows must be spaced some distance apart—a distance equal to 10 times the diameter of the blades. Side-to-side spacing is important too. Wind turbines must be spaced a distance equal to 3 diameters of the turbine blades. Spacing wind turbines more closely creates inefficiencies and could also damage the turbines. So, each 1.65-megawatt windmill (with 260-foot blade diameter) would need an area 2600 feet x 780 feet—that is, 200 thousand square feet (or approximately 45 acres). In short, producing electricity from wind takes a lot of space.

## Limits to Growth?

According to Travis Bradford (*Solar Revolution*, 2006), the dramatic growth in wind power in recent years occurs primarily because "at favorable sites, industrial-scale windmills have become cost effective compared to all other forms of electricity generation." Bradford adds,

> Large-scale use of wind to generate electricity is, at the moment, limited by the nature of the wind resource itself. Wind is intermittent, which causes the electricity that wind turbines provide to fluctuate, sometimes dramatically and unpredictably. If the wind speed is too low or too high to be useful or optimal, then a turbine is unusable for electric generation. As a result, wind turbines alone cannot be large-scale providers of electricity to the energy grid regardless of their cost-effectiveness and cannot reliably provide either peak power such as hydroelectric dams or direct solar or baseload power such as nuclear plants. If they are to be deployed as more than a small fraction of the electricity grid infrastructure, they must be coupled with backup generators or large-scale storage.

Wind can supply electricity only to places where the electrical grid is available for times when the wind is not blowing or when more electricity is required than the wind turbines can provide. In this respect even grouping wind farms together does not help much. A full-capacity, reliable, back-up energy source is required to supply energy when the wind is not blowing. This is a serious limitation. Both solar- and wind-power require the existence of a 100-percent back-up system. Said differently, solar- and wind-power can supplement existing power-generating systems, but they cannot replace those systems.

> Solar- and wind-power can supplement existing power-generating systems, but they cannot replace those systems.

## Costs of Wind Energy

The major cost of wind is depreciation of the equipment and interest on the investment.

The *installed capital cost per kilowatt* of wind power is about one-third the cost of solar power delivered from photovoltaic cells. Obviously, the "fuel" costs nothing. However, the *maintenance costs* for windmills over their entire lives, particularly as their size increases and as they age, are yet to be determined, but I suspect the costs will be significantly higher than current estimates. The sheer size and weight of modern, large windmills make maintenance inherently difficult and do not make maintenance costs easily predictable. For example, a 1.65-megawatt windmill has the hub of its blades about 225 feet from the ground, and the nacelle on top of the tower, behind the hub and the blades, is as big as a house, approximately 30' x 12' x 12', and weighs 55 tons. The hub and blades weigh another 55 tons. In addition, those magnificent spinning blades are vulnerable to severe weather conditions, high wind, large hail, lightning, ice, ultraviolet rays, bugs, and other atmospheric contaminants. For example, less than 1/32 of an inch of 'bug buildup' on the leading edge of the blades can reduce a windmill's power by as much as 25 percent, according to Howard Hayden in *The Solar Fraud*. Also, according to one windmill technician, windmills often take direct lightning hits so violent that the blade must be replaced.

There should be no direct government subsidies to the wind industry—or to the solar industry. Government subsidies are too political and too unreliable. The government should be involved in wind power only to facilitate and monitor fair practices. Future assistance can be provided by the energy surcharges I recommend elsewhere (see Chapter Sixteen). Simplicity is a great enabler.

Who should build wind farms? Utilities are the most logical industry to build wind farms. Utilities are the only group able to strike the right balance among intermittent sources of energy, a baseload, a back-up load, and a peakload. The next most logical group of wind-farm builders is oil and coal companies. These are integrated energy businesses and already work closely with utility companies. They are also generating loads of cash and need to find ways to secure their long-term futures.

Wind-farm owners with only a financial interest often do not understand the strategic significance of various energy sources. The wind farmers' lack of understanding can cause conflict between the wind-farm owner and the utility. I have nothing against small, individually owned wind farms or windmills. They have shown the way and make a lot of sense for specific applications. Many users already supplement their electrical needs with wind- or solar-generated electricity, which is a benefit since these energy sources reduce pollution and buy us time to completely abandon coal and other fossil fuels. Dedicated wind- or solar-power systems used directly by the end user always make sense whenever the back-up power system burns fossil fuel.

## WIND-ENERGY PRODUCTION WORLDWIDE

Some countries generate significant portions of their electrical energy from wind (see Figure 7.1). Germany generates about 10 percent and is growing rapidly. The German wind industry has already created about 35,000 jobs and is expected to create over 125,000 jobs in the next 5–10 years. Denmark generates more than 20 percent, and some say it will one day generate 50 percent.

**Figure 7.1. Top 15 and World Totals for Windpower Capacity (Megawatts)**

| Nation | 2005 | 2006 | 2007 |
|---|---|---|---|
| Germany | 18,415 | 20,622 | 22,247 |
| Spain | 10,028 | 11,615 | 15,145 |
| United States | 9,149 | 11,603 | 16,818 |
| India | 4,430 | 6,270 | 8,000 |
| Denmark (& Færoe Islands) | 3,136 | 3,140 | 3,143 |
| China | 1,260 | 2,604 | 6,050 |
| Italy | 1,718 | 2,123 | 2,726 |
| United Kingdom | 1,332 | 1,963 | 2,389 |
| Portugal | 1,022 | 1,716 | 2,150 |
| France | 757 | 1,567 | 2,454 |
| Netherlands | 1,219 | 1,560 | 1,746 |
| Canada | 683 | 1,459 | 1,846 |
| Japan | 1,061 | 1,394 | 1,538 |
| Austria | 819 | 965 | 982 |
| Australia | 708 | 817 | 824 |
| **WORLD** | **58,800** | **74,600** | **94,123** |

Sources: Global Wind Energy Council, 2007. Available at <www.gwec.net/uploads/media/07-02_PR_global_statistics_2006.pdf>.

European Wind Energy Association, 2007. Available at <www.ewra.org/fileadmin/eaea_documents/documents/publications/statistics/070129_wind_map_2006.pdf>.

Global Wind Energy Council, 2008. Available at <www.gwec.net/uploads/media/chartes08_EN_UPD_01.pdf>.

Denmark—a small country about one-half the area of South Carolina—is remarkably committed to wind, is ideally situated to take advantage of wind power, and requires only relatively short transmission lines. Denmark receives necessary back-up electrical power from neighboring Norway's hydropower. On a per capita basis Danes consume about one-half of the electricity consumed in North America, yet Denmark's per capita gross domestic product (GDP) is about the same as in the United States'. The world continues to learn from the Danes' experience and know-how as more and more wind power is installed around the world. It is no surprise that Vestas, the leading wind-turbine manufacturer, is Danish. (General Electric in the United States is also a large manufacturer of wind turbines. Other large, credible producers include Gamesa in Spain, Enercon in Germany, and Suzlon + REPower in India with assistance from Germany.)

All of the wind energy produced worldwide through 2006 is less than 1 percent of the total energy used in the United States.

Denmark, Sweden, and the United Kingdom plan to locate large wind farms offshore by 2020. Much bigger turbines, as large as 5 megawatts, are planned for these wind farms. I hope engineers accurately determined the maintenance costs in that hostile environment. Other important issues to address include noise, effects on birds and fish, maintenance, and the impact on ship navigation.

Some perspective: Denmark's wind-generated electricity accounts for about 25 percent of its total consumption of electricity, which is approximately 34 billion kilowatt hours of energy. Although wind in the United States generated about the same number of kilowatt hours in 2006, this energy accounted for less than 1 percent of the total U.S. *electricity* produced.

More perspective: Be vigilant when authors use numbers from other countries to prove a point. The scale and conditions are often very different, rendering comparisons almost useless. As stated, total Danish consumption of energy is less than 1 percent of the U.S. total.

We must build 3800 2-megawatt windmills every year for the next 17 years, at a cost of approximately $14 billion per year.

In fact all of the wind energy produced worldwide through 2006 is less than 1 percent of the *total energy* used in the United States.

## THE FUTURE of WIND POWER in the UNITED STATES

The Midwest area of the United States, from North Dakota to Texas, is sometimes called the Saudi Arabia of wind. North Dakota alone has enough potential wind energy to supply over 30 percent of the total electricity consumed in the United States. (For current installed wind capacity in the United States, see Figure 7.2.) Windlogics, a company in St. Paul, Minnesota, collected wind data for just about every square foot on the face of the earth to help companies select the best sites for wind generators. Determining the site of a turbine can be critical since sometimes moving a turbine just 1 mile can increase the energy output by 50–100 percent.

**Figure 7.2. Installed Windpower Capacity in the United States, January 2008**

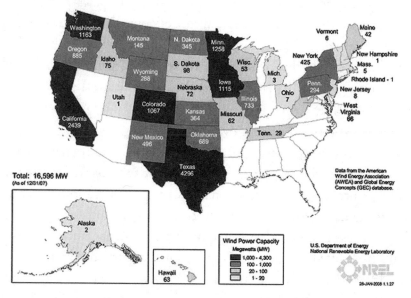

Source: National Renewable Energy Laboratory. Available at <http://www.eere.energy.gov/windandhydro/windpoweringamerica/wind_installed_capacity.asp>.

Let's say that half of the *new* electrical energy needed in the United States by 2025 will be produced from wind. Assume direct solar energy provides the other half. Experts anticipate a 20 percent increase in energy demand by 2025. If wind and solar together can supply this increase, then U.S. citizens should rejoice since such production will be a Herculean task. If wind power and solar energy meet this need, together they will be supplying in 2025 about the same percentage of power (20 percent) that the nuclear industry supplies today (22 percent). How might the wind industry pull off this feat?

## A Plan or a Hope?

We can project U.S. energy needs based on expected growth in the population and the economy. Based on those projections, the United States needs to build 65 thousand 2-megawatt windmills to supply *half of the anticipated increase* in our energy needs by 2025. (Note: The increased energy does not take into account the additional energy necessary for an all-electric transportation system or an all-electric economy.) In other words, we must build 3800 2-megawatt windmills every year for the next 17 years, at a cost of approximately $14 billion per year. Very do-able, assuming producers can build the turbines fast enough. If we build these windmills, then by 2025 wind energy would be generating about 10 percent of the then-total electrical needs in the United States.

Some experts think that over the next 50–75 years the United States could generate up to 50 percent of its electrical energy from wind power and direct solar energy. If they are to be backed-up by fossil fuel plants, then this prediction makes some sense. However, if we are already using fast neutron reactors, then wind and solar systems, which are less clean and less economical than modern nuclear power, would be limited to specialty applications at particularly attractive sites. However, I speculate that in 100 years we could slowly begin generating all electrical energy from very efficient solar cells coupled with high-capacity batteries.

In the meantime let's try to satisfy 10 percent of our energy needs from wind and 10 percent from direct solar, particularly where circumstances make these alternatives most attractive. The effort will create new knowledge and innovations and likely lower costs. Thereafter experience will dictate whether more wind or solar energy makes sense. Besides the technology and cost, timing will be a factor.

I've actually seen no projections with solid quantification and a reasonable timeline for the role wind could play in a clean energy future—what you hear from the wind-power industry is mere speculation, more a dream than a sound plan.

## Enhance Efficiency by Separating Hydrogen?

A windmill produces energy only about 35 percent of the time. That is, its *capacity factor* or *service factor* is 35 percent. When wind energy is not needed by the grid because of temporary low demand, the energy is wasted since there is no available, cost-effective system for storing large amounts of energy.

In cases when the grid does not need wind energy or solar energy, the grid could direct this energy to one or more electrolysis plants, strategically located to separate water into hydrogen and oxygen. Even though separating water by electrolysis is not particularly efficient, it certainly would be if the electrical energy is free. Subsequently, the hydrogen could be used to provide energy when the wind is not blowing and when the demand is more than the grid can provide. All utilities know how to manage this. Recovering this presently wasted output could improve the overall efficiency of wind energy. However, a full back-up energy-producing system is still necessary for when the wind is not blowing and all the stored energy is depleted.

## Gridlock

To have any hope of meeting the goal of supplying half of the expected increase in needed power, both wind power and solar energy will require massive, expensive extensions and modifications of the existing U.S. electrical grid. It's no secret that the existing electrical grid has been neglected and will require billions to modernize. Also, high-intensity wind locations and high-intensity solar locations are far from major population centers, so may require the addition of transmission lines and towers. Further, grid-managing jurisdictions appear to have no effective coordinating body. Consequently when costs, technologies, or purposes are not agreed upon, nothing gets done. This must change if solar energy and wind power are to make much of a contribution to grid electricity. Fixing the grid has all the earmarks of a legal and political nightmare, so don't count on this getting taken care of any time soon.

To illustrate that we are adrift and without direction, I refer you to a 2007 report by the U.S. National Research Council. The report criticizes the "lack of any truly coordinated planning" in the growth of wind farms across the United States and calls on all levels of government to pay more attention to the effects of turbines on wildlife and scenic landscapes. The NRC report goes on to say that from 2000 to 2006, U.S. wind capacity quadrupled. By 2020 the report predicts wind power could offset 4.5 percent of the planet-warming carbon dioxide that U.S. entities would otherwise spew into the atmosphere. However, the anticipated wind energy in 2020 would contribute only one-third of the *increase* in energy needed between now and 2020, based on projected population growth alone. This much wind energy would do nothing to reduce the total amount of pollutants going into the environment now. Therefore, even with dramatic increases in the production of robust, clean, renewable wind energy, the total amount of pollutants emitted in 2020 will be greater than today. Some believe that wind can be the final energy solution, but it cannot and will not.

> The existing electrical grid has been neglected and will require billions of dollars to modernize.

## DON'T GET BLOWN AWAY

You and I should vigorously support wind power, particularly if the alternative is the continued use of fossil fuels. The more energy derived from wind, the less pollution and the less reliance on foreign sources. At the same time, we must be realistic and not be led down impractical paths that cause us to disregard other solutions. Wind has advantages and huge potential, but it also has some significant limitations. Our industrial and political leaders should have a broad understanding of the facts before formulating policy.

## Limits to Wind Power

▷ **Extraction.** Windmills extract only a small portion of the energy from the wind that hits the turbine blades.

▷ **Spacing and Space.** Each 2-megawatt turbine needs about 45 acres. To generate 10 percent of America's energy from wind would require an area 100 miles by 70 miles.

▷ **Transmission Losses.** How do we get North Dakota's wind energy to New York? The energy lost per 100 miles of transmission can reach 7 percent. At present, transmission losses nationally cost about 2.5¢ per kilowatt hour—often 50 percent or more of the total cost of generating electricity. Connection costs and transmission costs often determine whether introducing a renewable energy source makes sense.

▷ **Variable Power.** Some utility engineers describe wind power as the lowest-quality power available, because it is intermittent and fluctuates, and depends on the velocity of the wind. Utilities can compensate for fluctuations in wind power only when the wind supplies less than 10 percent of the power going to a large grid. I am sure engineers will learn to manage this problem more effectively in the future.

▷ **Harm to Wildlife.** Yes, windmills kill some birds and bats, but so do BB guns, slingshots, cars, buildings, exterminators, and pollution. Get real. Oil and gas extraction kills 1–2 million birds per year. According to a study by Curey and Kirlenger, compiled from governmental organizations and environmental groups, more than 100 million birds annually smash into windows in buildings; cars and trucks kill another 50 million each year; and cats kill about 100 million birds annually, 7 million in Wisconsin alone. If you want to save birds, then stop driving, keep your windows dirty, and keep the cat indoors. I do not mean to trivialize the bird and bat problem, but please consider the alternative—using up the world's precious resources and spewing toxic pollutants kills birds, bats, and humans, too. By comparison, windmills when supplying 10 percent of the nations's electrical energy will kill less than 10 percent of the birds killed by the oil and gas industry.

▶ **Severe Weather.** Lightning frequently strikes windmills, causing serious damage. Windmills are also subject to other dangerous, severe-weather conditions, such as hurricanes, cyclones, and tornados.

▶ **Build up of Bugs and Ice.** Such buildup can lower efficiency by approximately 30 percent and the blades are costly to clean.

▶ **Maintenance.** Over the life of the windmill, maintenance costs may be much higher than original estimates. (See Figure 7.3.)

### Figure 7.3. Maintenance on a Windmill

Source: Sandia National Laboratories at <www.nrel.gov/data/pix/searchpix.html>.

## More Misinformation

On a recent ski trip to Beaver Creek, Colorado, I did a double take when I saw signs stating that 100 percent of the town's electricity was derived from the wind. It couldn't be true unless the wind blew continuously in a specified velocity range. It doesn't. Not even in Beaver Creek. In this wealthy community people pay an extra 1–2¢ per kilowatt hour to have clean wind energy—to do their part to save the earth. Upon further investigation I learned the residents actually bought wind *credits* so the local utility could purchase more green energy. In reality the town obtained only 7 percent of its electrical energy from renewable sources, with about 70 percent (5 percent of total) of this energy coming from hydropower. It is absurd to claim 100 percent wind power. You still get what's on the grid—period. This kind of false information only confuses and distorts the town's true energy picture, and this from a provider called Holy Cross Utility. To their credit, residents were trying to do the right thing—but be careful. As far as I can tell, there is no accounting for exactly what their extra payments buy, or any determination of whether the payments yield a net increase in wind energy any place in the United States.

## BOTTOM LINE

> The United States should do everything possible to generate 10 percent of the nation's energy needs from wind power as soon as possible—certainly by 2040. This is no trivial task, since this amount of electrical energy is more than the total electrical energy consumed in Denmark, Norway, Sweden, and Finland combined. This goal should not be deterred by lack of funding.

> A large amount of money should be allocated immediately to support and overwhelm research efforts directed to the development of recyclable, high-energy-density batteries

for use in wind, solar, transportation, and manufacturing industries. The United States and the world need such batteries, which will enable many advances in many dependent industries.

*Chapter 8*

---

# BIOFUELS

## PART OF THE SOLUTION

I was amazed to learn that in 2006 the United States produced more energy from biomass than from all of the 75,000 U.S. hydro-electric power plants combined.

Six percent of U.S. energy is renewable. Of the renewable energy, biomass and hydro-electric power account for 92 percent. Geothermal is 6 percent, wind 1 percent, and solar one-half percent. Farmers, no longer raising only food, are beginning to play a significant role in producing fuel and bioenergy. The biofuel industry is here to stay, and it should have the benefit of our total support.

Biofuels derived from organic matter (biomass) are efficient fuel sources, so make a lot of sense. Biofuels emit far less carbon dioxide than gasoline does, because growth of the organic matter removes carbon dioxide from the atmosphere. The production of biofuels, dramatically ahead of growth projections, is on its way to becoming a $100-billion industry producing corn ethanol and soybean biodiesel now and cellulosic ethanol and algae-based biodiesel in the near future.

Coupled with plug-in hybrids or all-electric vehicles, ethanol and biodiesel will likely supply the mostly carbon-free liquid fuel required for transportation, industry, residences, and many other uses.

At present, the 2 most important biofuels are ethanol from corn and biodiesel from various crops, mainly soybeans in the United States.

## ETHANOL

Ethanol fuel is concentrated common beverage alcohol rendered unfit to drink. Humans made alcohol before recorded history. It likely started with an accidental fermentation and an accidental sip. The rest is history. It took centuries to figure out that a living micro-bug did all the work of converting sugar to alcohol. It took Louis Pasteur and the microscope to find out what was happening. A fungus (yeast) grows in a sugary liquid, producing alcohol and carbon dioxide.

In 1908 Henry Ford assumed ethanol would fuel his cars. Ford missed by about a century. What can ethanol do for us today?

- Ethanol lowers the levels of toxic ozone-forming pollutants and greenhouse gases.
- Ethanol doesn't leave gummy deposits, thus helping fuel systems maintain optimum performance.
- Ethanol extends gasoline supplies.
- Ethanol replaces MTBE (methyl tertiary butyl ether) as an octane-enhancer.
- Ethanol creates jobs in the United States.
- Ethanol improves the negative U.S. balance of payments by reducing imports of oil and supports the value of the dollar.
- Ethanol helps farmers get fairer prices for their products.

### Ethanol Production

Processors begin producing ethanol by grinding organic matter, typically corn, to coarse flour that is then combined with water and enzymes. The enzymes convert starch to sugar, creating a mash that is cooked and sterilized. After cooling, yeast is mixed with the mash to ferment the sugars. Finally the fermented mash is distilled to extract the ethanol. The high-grade carbon dioxide, which the fermentation produces, can be liquefied and sold.

The mash remaining after the ethanol is extracted is either pressed through a screen or sent to a centrifuge to remove the liquid. The liquid is recycled back into the cooking system or sold as livestock feed. The spent grains are sold as livestock feed called distillers' wet grains or distillers' dried grains (DDGs). DDGs are an important, high-energy, high-protein animal feed.

## Corn Ethanol

There are 131 ethanol plants in production in the United States and 61 under construction, primarily in the Corn Belt. Production of ethanol from corn was predicted to reach 6.5 billion gallons in 2007, equivalent to about 4.3 billion gallons of gasoline. Such production is a remarkable achievement, but still only 2 percent of total U.S. transportation fuel. (The United States annually consumes more than 140 billion gallons of gasoline and about 60 billion gallons of diesel.)

Present ethanol production consumes about 21 percent of the United States corn crop. Thus, if the U.S. devoted the *entire* corn crop to ethanol, the fuel would satisfy only about 10 percent of our transportation needs, and less than 7 percent in 30 years.

Since converting corn to fuel will obviously not bring us much closer to the goal of energy independence, and since global population will grow by several billion in the next 30 years, is producing corn-based ethanol worth the effort? Yes. Experts believe that corn production will double in 30 years and that one-third of the total U.S. corn harvest can be diverted to ethanol production without disrupting the food supply. Also, in 30 years

> If the United States devoted the entire corn crop to ethanol, the fuel would satisfy only about 10 percent of U.S. transportation needs.

I am quite certain that hybrid plug-ins and all-electric vehicles will be common and much less transportation fuel will be required.

### How Much Energy Does it Take to Produce Corn Ethanol?

Many believe that producing corn ethanol uses more energy than the corn ethanol contains. This deficit is called a *negative energy balance*. Virtually every study shows a *positive* energy balance of 20–30 thousand BTU per gallon. However, one frequently cited study—by Dr. David Pimentel, an emeritus professor of ecology and agriculture at Cornell University, and Tad W. Patzek, a professor of civil and environmental engineering at the University of California at Berkeley—reports a net negative energy balance of *minus* 20–35 thousand BTU per gallon, an astounding difference.

Dr. Pimentel is a highly respected professor. However, when studying his work, it seems he confuses capital costs with operating costs. And even if it does take more energy to deliver ethanol than the ethanol contains at the pump, the same is true for gasoline. It takes about 1.25 gallons of gasoline equivalence to deliver 1 gallon of usable gasoline to the gas pump. This ratio is even worse for coal. And ethanol pollutes less. Corn ethanol E85 reduces greenhouse gas emissions by 20 percent when compared to gasoline. Other than Pimentel and Patzek, most reports show that ethanol has a much better energy in/energy out ratio than gasoline. In the meantime, it reduces U.S. dependence on foreign oil, reduces U.S. negative balance of payments, and supports the value of the dollar—facts that should also be factored into any cost comparison between ethanol and gasoline.

### Improving Ethanol Production

Since 1980, processors have reduced the energy needed to produce ethanol by over 40 percent. Significant opportunities remain to further cut costs. First, an oil can be extracted from the high-protein residue (DDGs) to produce 5–10 percent more fuel in the form of biodiesel. In the process, according to the Center for Energy and Environment, the quality and selling price of the high-protein residue

(DDG) would also improve. Second, clean electricity could be produced from currently wasted heat, thereby reducing total costs. Third, experts anticipate that costs will be cut through better recycling and management of water. The amount of water used to produce ethanol is very high, but will likely be cut in half, from 3 gallons to 1.5 gallons for every gallon of ethanol produced. Still, the amount of water required to produce ethanol is dramatically *less* than the water required to refine oil—about 44 gallons of water per gallon of crude oil, according to the EPA.

> The amount of water required to produce ethanol is dramatically less than the water required to refine oil.

Sugar cane is an ideal crop for making ethanol because it simplifies processing. Brazilian processors produce alcohol from sugar cane for about 32¢ per gallon. The cost for sugar cane producers in Florida is about 55¢ per gallon. Other crops can be used, but none are available in sufficient quantities in the United States to make much of a difference (see Figure 8.1).

## Figure 8.1. Commercial Average Yield of 200-Proof Ethanol Alcohol

| Material | Gallons per Bushel |
|---|---|
| Wheat | 2.56 |
| Corn or Milo | 2.34 |
| Rye | 2.19 |
| Buckwheat | 1.99 |
| Barley | 1.89 |
| Oats | 1.01 |
| Sweet potatoes | .93 |
| Potatoes | .68 |
| Jerusalem Artichokes | .59 |

Source: *The Alcohol Fuel Handbook* by Lynn Ellen Doxon.

All in all, costs are falling and much progress has been made in producing corn ethanol. I believe cellulosic ethanol—ethanol produced from biomass, such as grasses and trees—is an even better choice, but if we had not developed a corn-ethanol industry, we would not be in a position to exploit these new possibilities.

### Ethanol Subsidies

Subsidies for the production of ethanol have been substantial, highly criticized, and significant for the growth of this industry. The industry still gets a subsidy, but just about all of it now goes to the "blender" (oil companies). I found much that has been written about these subsidies is incorrect. Richard Conniff in the *Smithsonian* reports that direct subsidies to corn producers amounted to $9 billion in 2005. This is ancient history. In fact, in 2006 and 2007 the U.S. government collected much more money from the ethanol industry than it gave in subsidies, and, as one producer suggested, it is now time to end all subsidies.

As a result of corn being used to produce corn ethanol, the price of corn has risen above its support level, saving taxpayers over $6 billion per year. In addition, in 2006 the ethanol industry paid $5 billion in federal, state, and local taxes on revenues of over $60 billion, while also creating more than 150,000 jobs throughout the economy, not to mention the positive effects on U.S. balance of payments. Negative balance of payments equals about $1 billion per day for oil imports and suppresses the value of the dollar. Of course, the economics are more complicated since as corn prices rise so do the costs for corn-based foods.

The subsidies amount to 51¢ per gallon, yet almost all of it goes to ethanol "blenders," which are most often oil companies. The petroleum industry didn't lobby for the subsidies, but it pockets them. Congress created this subsidy (gift) to encourage a larger ethanol market, figuring the oil industry needed a financial incentive to accept ethanol, the new kid on the block. Here's the kicker: Blending costs the oil companies nothing additional. They have to blend

This subsidy is the legendary Robin Hood story in reverse. In 2006 the top 4 oil companies earned $107 billion dollars.

something into gasoline to improve the octane. The current octane enhancer—MTBE (methyl tertiary butyl ether)—was introduced in 1992 to replace lead, but is now being phased out because it too causes cancer. Ethanol to the rescue. About 14 billion gallons of ethanol will be needed annually to replace the banned MTBE. Blending 10 percent ethanol into gasoline raises an 80-octane gasoline to 87.

Isn't it ironic that oil companies now need the ethanol industry, the very industry the oil companies tried to discredit? Apparently, oil companies believe the ethanol industry should recognize its deferential, subordinate role. David Kiley in *Business Week* reports, "Despite collecting billions for blending small amounts of ethanol with gas[oline], oil companies seem determined to fight the spread of E85, a fuel that is 85 percent ethanol and 15 percent gas[oline]." The article goes on to say, "While oil reps say they aren't anti-ethanol, they are candid about disliking E85." Kiley quotes Al Mannato of the American Petroleum Institute (API), the chief trade group for oil and natural-gas companies: "We think [ethanol] makes an effective additive to gasoline but that it doesn't work well as an alternative fuel. And we don't think the marketplace wants E85." Who cares what he thinks? It is certain that the market does not like the prospects of depletion, higher prices, and pollution caused by oil products. Is the problem that the market doesn't want ethanol, or is it that oil companies don't want it? Of the 179,000 fuel pumps in the United States only about 1000 pump E85. Almost none of these pumps are at oil-company-owned gasoline stations. Apparently, as *Business Week* put it, "they don't want their brand assaulted by someone else's product." The government should simply mandate the presence of biofuels at service stations. Some ethanol plants have suspended operations for a time because their product is not made available at most stations.

## A Rant About Oil Companies and Politics as Usual

So, the oil industry gets a government subsidy—over $3 billion dollars in 2006—to blend ethanol, exactly what it would do anyway. This is the legendary Robin Hood story in reverse. In 2006 the top 4 oil companies earned $107 billion dollars. I could not identify all oil industry subsidies because they are hidden behind terms like *depletion allowance, tax credits, research-and-development supports,* and *blending fees,* but the total seems certain to reach tens of billions of dollars per year. That's at least 10 percent of earnings. Doesn't it frost you to know your hard-earned tax dollars contribute through subsidies to their fat bottom line? Are the legislators showing their gratitude for the campaign contributions received from oil companies?

Instead of fighting new fuels, oil companies should become energy companies and build ethanol plants and invest in wind, solar, and nuclear power. They have the infrastructure, talent, and money. They dabble in renewable energy, but I suspect much of it is window dressing. Instead, they have boat loads of money to convince us that fossil fuels remain the answer. Although powerful now, I think oil companies will fade into obscurity if they don't change their ways.

## Cellulosic Ethanol

Corn and sugar crops represent only a small fraction of biomass that can produce ethanol. Several technologies can produce ethanol from other forms of biomass, such as grasses, trees, forestry residue, and plant stalks, as well as industrial and domestic waste, and even municipal solid waste.

Cellulose, the matter that gives plants their structure, is made of sugars. Ethanol can be produced when a plant's cellulosic sugar is broken down into simpler fermentable sugars. Low-value plant material such as corn stalks, grasses, fast-growing trees, sawdust, waste paper, and paper-mill waste can be used to produce ethanol, except the processes are currently more expensive than producing ethanol from high-value plants such as corn. Plants such as switch grass and fast-growing trees can be grown on marginal or degraded land presently unsuitable for food crops.

Producing cellulosic ethanol requires a few more processing steps than producing corn ethanol. However, plants processing corn ethanol can be converted to cellulosic-ethanol production for approximately 25 percent of the original plant's cost. This is good news since presently invested capital in corn-ethanol plants need not be abandoned when and if a switch is made to cellulose-based ethanol. On the positive side, according to Conniff in *Smithsonian*, burning 1 gallon of cellulosic ethanol promises to cut greenhouse gas emissions by at least 80 percent compared to burning 1 gallon of gasoline, yet most sources conclude that corn ethanol alone reduces greenhouse gases by only about 20 percent. Cellulosic ethanol also has a more favorable energy in/energy out ratio than corn ethanol. However, some sources, such as the *Smithsonian* article mentioned above, argue that greenhouse gas emissions actually increase if the ethanol processing plants get the energy for fermentation from coal-powered plants. The *Smithsonian* calculation fails to mention that the crop requires no fertilizer, no planting, no cultivating, no pesticides, and no herbicides, all requiring considerable energy. If we factor in the absence of that expended energy, then ethanol production from cellulose reduces greenhouse gas emissions.

Cellulosic ethanol is now produced for about $2.15 per gallon, a figure competitive with gasoline from oil costing $135 per barrel. In 5 years the cost of cellulosic ethanol is projected to fall to $1.07 per gallon, a price comparable to gasoline processed from oil costing less than $70 per barrel. If the United States starts building processing plants now, then the country should have cellulosic ethanol, per the industry's plan, for about 60¢ per gallon in less than 10 years (see Figure 8.2).

Cellulosic biomass could replace 30–50 percent of the petroleum used for U.S. transportation.

That price is comparable to processing gasoline from oil costing $36 per barrel—a great bargain since oil prices were flirting with $100 per barrel in November 2007, reached a history-making high of $117 per barrel in April 2008, and will likely rise. Once in the game, so to speak, with commercial plants, costs *always* improve.

Industry must allocate money for research and development, and government must offer subsidies—up to $1 per gallon derived from a gasoline surcharge of 50¢ per gallon, yielding $70 billion per year.

That's a lot of money, but it pays for sustainability, carbon reduction, and a grand step toward energy independence. It is imperative to get cellulose ethanol off the ground *now*. Facilities for producing cellulosic ethanol should be built and subsidized until the production is competitive with gasoline production, but no longer. The products become competitive when the costs per BTU for cellulosic ethanol and gasoline are the same. It won't be long.

### Figure 8.2. Costs of Ethanol from Cellulosic Biomass (2006 U.S. Dollars)

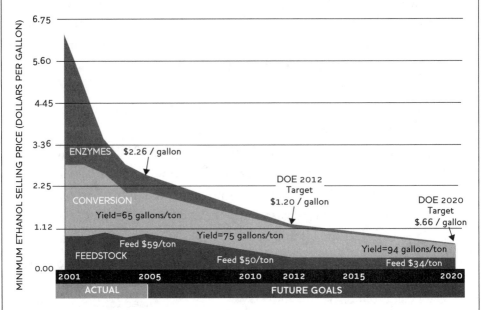

Source: U.S. Department of Energy, Biomass Program Data.
Actual cost in 2007 was $2.15 per gallon.

### *Which Feed Stocks (Plants) to Use?*

I believe a combination of switch grasses, hybrid poplars, cotton-woods, and corn stalks will be the predominant feed stocks for producing ethanol in the future. Other crops will enter this mix once the industry is established. For deciding which plants to use, the following must be considered: gallons per acre (energy density), feed stock cultivation, growing cycles, fertilizer requirements, resistance to infection, and location. Other than available wood waste, grasses seem the best choice. Grasses can be grown on a 10-year cycle and can be harvested the first or second year. Trees require a 6–20 year rotation and can be harvested only in years 4–10.

Switch grass, in combination with other grasses, has the highest potential for ethanol production. While an acre of corn yields about 400 gallons of ethanol, some grasses will yield more than 1000 gallons of ethanol per acre. If the grasses are grown in complementary combinations, then the yield could double. Switch grass grows well just about anywhere, grows quickly, needs little water, requires no insecticides or fertilizers, prevents soil erosion, and can be harvested repeatedly because it continually restores nutrients to the soil. For an idea of what a good, dense biomass crop would look like, see Figure 8.3. Wildlife habitat? Guaranteed.

**Figure 8.3. Miscanthus (Grasses)—A Single Season's Growth in Illinois**

Source: Department of Energy.

### More Good News

The total carbon emitted from producing cellulosic ethanol and burning the produced fuel is often less than the carbon taken from the air by the biomass used to produce the ethanol. For some grass combinations the ratio of $CO_2$ emitted to $CO_2$ absorbed is very favorable, and could be better than 2 to 1 positive. Just ask David Tilman, Jason Hill, and Clarence Lehman, each from the University of Minnesota. Their article on ethanol from grassland biomass appeared in *Science*:

> Biofuels derived from low-input high-diversity (LIHD) mixtures of native grassland perennials can provide more usable energy, greater greenhouse gas reduction, and less agrichemical pollution per hectare than can corn grain ethanol or soybean biodiesel. High-diversity grasslands had increasingly higher bio-energy yields that were 238 percent greater than monoculture yields after a decade. *LIHD biofuels are carbon negative because net ecosystem carbon dioxide sequestration (4.4 mega gram per hectare per year of carbon dioxide in soil and roots) exceeds fossil carbon dioxide release during bio-fuel production (0.32 mega gram per hectare per year).* Moreover, LIHD bio-fuels can be produced on agriculturally degraded lands and thus need to neither displace food production nor cause loss of biodiversity via habitat destruction (emphasis added).

> For some grass combinations the ratio of $CO_2$ emitted to $CO_2$ absorbed is very favorable, and could be better than 2 to 1 positive.

## Ethanol's Potential

How much ethanol from biomass can we produce? Numbers vary greatly, but the best practical estimates suggest that we can grow up to 1.3 billion tons of dry cellulosic biomass, which could replace 30–50 percent of the petroleum the United States presently uses for transportation. With predicted population growth in 30 years, biomass ethanol's contribution falls to 20–35 percent. Of this cellulosic alcohol, energy crops (corn, grasses, and hybrid trees) would yield about 30 percent, agricultural residues would account for about 35 percent, and municipal waste and lesser sources account for the remaining 35 percent.

However, a more complete solution stares at us. *If* the U.S. public cuts gasoline consumption 60–90 percent by using hybrid plug-ins and all-electric cars *and* uses cellulosic ethanol and biodiesel to fuel these vehicles, *then* the United States would become essentially energy independent for transportation. The country must do this in 30 years. If the United States adds windmills, solar cell systems, and fast neutron reactors, then the United States becomes totally energy independent. If the rest of the world follows a similar path, future generations will live in a more peaceful, environmentally clean world. A naïve pipe dream? Don't believe it—it can be done.

We must wisely choose the balance of fuels, energy sources, and the like, and then change as experience dictates.

Let's cut to the chase: The United States must reduce transportation-petroleum needs to zero by the year 2040. Two changes are necessary to succeed. Ninety percent of all vehicles on the road must become hybrid plug-ins or all-electric, and ethanol production must reach 50 billion gallons per year. Impossible? We put people on the moon in 10 years, didn't we? Americans love a challenge, and this one is worthy of our mettle.

> Ninety percent of all vehicles on the road must become hybrid plug-ins or all-electric, and ethanol production must reach 50 billion gallons per year.

Our need for liquid transportation fuel decreases in proportion to the conversion to plug-in hybrids and all-electric vehicles. When all cars, light trucks, and some heavier vehicles become hybrid plug-ins or all-electric, energy crops will serve other purposes in a world of rapid population growth.

- Energy crops could replace corn as cattle feed, freeing up the corn for human consumption.

- Cellulosic power plants could produce electricity with a net of zero carbon dioxide emissions.

- Energy crops, instead of petroleum, can be used to produce plastics and other chemicals.

## BIODIESEL

Biodiesel is a fuel derived from biological sources, such as vegetable oils, animal fats, and even recycled restaurant greases and oils. The fuel, which can be used in its pure form (B100) or in combination with petroleum-based diesel in any ratio, requires at most only minor engine modifications. Although biodiesel's energy content is about 93 percent that of petroleum diesel, this shortfall is offset by biodiesel's better lubricating properties, resulting in more engine efficiency and longer engine life. Also, biodiesel is biodegradable.

Biodiesel is better for the environment than petroleum-based diesel and other fossil fuels. Compared to petroleum-based biodiesel, pure biodiesel spews about half the particulate matter and carbon monoxide, and it emits even lesser amounts of other toxic pollutants. Sulfur emissions, a major source of acid rain, is essentially eliminated. A joint study by the U.S. departments of Energy and Agriculture concludes that biodiesel fuel, due to its closed carbon cycle, reduces net carbon dioxide emissions by 78 percent compared to petroleum diesel. The $CO_2$ released when biodiesel burns is re-absorbed by (recycled back to) the next crop of plants to be harvested for processing.

Also, biodiesel has a very favorable energy ratio. The joint study reports that for every unit of fossil energy used to make biodiesel, 3.2 units of energy are gained.

The production of biodiesel mirrors ethanol, but on a smaller scale. Like ethanol, biodiesel's feed stock is often a farm food crop. Many crops can be used, but rapeseed and soybeans are most common. Soybean oil accounts for about 90 percent of the biodiesel production in the United States. Although the food-versus-fuel debate is the same, soybean oil has been in surplus (see Figure 8.4).

Making biodiesel fuel is a fairly simple process. The vegetable oil is removed from the source plant, and the oil then undergoes a process called *transesterification*, whereby the raw vegetable oil reacts with an alcohol: 100 units of vegetable oil + 10 units of alcohol produces approximately 10 units of glycerin + 100 units of diesel fuel.

Producing ethanol from starchy feed stocks like corn is considerably more complex.

## Biodiesel's History

The transesterification process was conducted in labs as early as 1853, many years before the first diesel engine became functional. Rudolf Diesel's engine ran on its own power for the first time in Augsburg, Germany, in 1893. Diesel later demonstrated his revised engine in 1900 at the World Fair in Paris, France, where he received the *Grand Prix* (highest prize).

### Figure 8.4. Vegetable Oil Yields
### (The biodiesel yield = oil yield x 0.8 approximately)

| Crop | Gallons Per Acre | Crop | Gallons Per Acre |
|---|---|---|---|
| algae | 5,000* | rice | 88 |
| oil palm | 635 | safflower | 83 |
| coconut | 287 | sesame | 74 |
| avocado | 282 | camelina | 62 |
| brazil nuts | 255 | mustard seed | 61 |
| castor beans | 151 | euphorbia | 56 |
| olives | 129 | hazelnuts | 51 |
| rapeseed | 127 | linseed (flax) | 51 |
| opium poppy | 124 | soybean | 48 |
| peanuts | 113 | hemp | 39 |
| sunflowers | 102 | cotton | 35 |

* Research conducted by the National Renewable Energy Laboratory indicates 15,000 gallons per acre is theoretically possible.

Source: Journey to Forever, an environmentally minded non-profit, non-governmental organization. Available at: <http://journeytoforever.org/biodiesel_yield.html>.

Diesel believed one of the great advantages of his engine was its ability to run on biomass fuel. His original engine was powered by peanut oil—a biofuel, though not *biodiesel*, since it was not transesterified. In a 1912 speech Diesel said, "The use of vegetable oils for engine fuels may seem insignificant today, but such oils may become, in the course of time, as important as petroleum and the coal-tar products of the present time." Very prophetic. During the 1920s the petroleum industry came to dominate fuel markets because its fuel was much cheaper to produce than biomass alternatives. The result was a near elimination of the biomass fuel-production infrastructure. Recently, rising concerns over environmental impacts and decreasing price differences between the fuels make biodiesel a competitive alternative.

Throughout the 1990s, biodiesel plants opened throughout the world and in many European countries. By 2000, 21 countries had commercial biodiesel projects. France launched local production of biodiesel fuel (called *diester*) from rapeseed oil. This fuel is mixed into regular diesel fuel to a level of 5 percent, and into the diesel fuel used by some fleet vehicles to a level of 30 percent. Many service stations across Europe offer 100 percent biodiesel. In 2004 Europe produced about 10 times more biodiesel than the United States.

In the United States in 2005, Minnesota became the first state to mandate that all diesel fuel sold in the state contain some biodiesel. Minnesota requires at least a 2-percent mixture.

In the United States today, more than 700 truck fleets, some city buses, and some government fleets use some biodiesel. Across the country 165 plants produce biodiesel, with 80 more under construction. Although the U.S. production of biodiesel in 2007 approached 325 million gallons, total U.S. consumption of diesel is 60 billion gallons per year. Biodiesel therefore contributes less than 1 percent of the total. If the United States converted 100 percent of its production of soybean oil to biodiesel fuel, the result could replace only about 7 percent of the total diesel fuel consumed today in the United States. Not much right now, but the total could significantly increase as new technologies emerge.

As with the ethanol industry, the biodiesel industry has a plan. By 2015 the biodiesel industry plans to replace 5 percent of the *on-road* consumption of petroleum diesel. Since on-road consumption is approximately 38 billion gallons per year, biodiesel production must increase to 1.9 billion gallons annually. With present and planned production capacity, this plan is attainable. The plan assumes grain commodity prices will cooperate. This plan anticipates biodiesel derived only from food crops.

In the near future, on-going research and development will certainly lead to the production of biodiesel from algae and other non-food crops.

> If the government wants to promote the use of biofuels, then mandate their use. Governments should quit playing money games that distort the realities of an industry.

## Algae Biodiesel

Ethanol production will be greatly enhanced when the conversion of cellulosic biomass to ethanol becomes an established industry. Biodiesel has its own next-generation counterpart with algae, which could greatly enhance the production of biodiesel. Neither process would require the use of food crops.

Theoretically, *some algaes could produce up to 15,000 gallons of biodiesel per acre.* Although such production is probably impossible, even 1000–5000 gallons per acre represents a *huge* advantage over soybeans or any other food crop as biodiesel sources. Open ponds and photo-bioreactors are being developed to exploit this opportunity. Some knowledgeable industry people think algae biodiesel could replace all transportation fuels. Let's get to 10 percent first, and let's do it fast. To this end, this industry should receive the support it needs to get established. This support could come from the recommended surcharges on all energy (see last chapter).

## Biodiesel Subsidies

The bulk of any biodiesel subsidy goes to the blender, usually an oil company. The blender receives a tax credit of $1 for biodiesel made from new crops and 50¢ for biodiesel made from recycled oil. Once again, this gift from the government to the oil companies is in thanks for accepting biofuels in their fuel mix. The tax credit for the blender should be eliminated immediately. If the government wants to promote the use of biofuels, then mandate their use. Governments should quit playing money games that distort the realities of an industry.

The biodiesel industry, as with the ethanol industry, recently paid more in taxes in a single year than it received in subsidies. And the industry creates jobs (40,000 by 2015). Said differently, the subsidies helped generate a net gain in government revenue and prevented over $10 billion in oil imports. Last, since soybean prices are rising, some farmers no longer need crop subsidies, resulting in further savings. All in all, the biodiesel and ethanol industries contribute significantly to the U.S. economy.

## SUBSIDIES 101

The word *subsidy* is as painful to some ears as *taxes* and *welfare*. The political and emotional reactions are just as strong. "Subsidies are government hand outs." "That's my money the government gives away." "Why should a business get a break? I don't get one." "I want free markets, not the government meddling in the market." Let's get away from broad concerns and turn to a practical question: If a specific subsidy benefits society and consumers as a whole, then is it okay? Yes, I think so. Indeed, I think subsidies to producers of ethanol and biodiesel are great buys for the U.S. public. Great for the pocketbook. Great for the environment. Great for energy independence, balance of payments, value of the dollar, and national security. Great for our kids and the future.

Here's why. I think the facts will astound you.

First, if you dislike subsidies in general, then you should hate the subsidies received by oil companies and the subsidized, artificially low price of gasoline. If gasoline were priced according to a truly free market, then it would cost much more per gallon. If gasoline were priced according to a truly free market *and* if the price included social costs (such as the costs of pollution on health, for example), then the price would be much, much higher per gallon. In effect, a subsidy (subsidized costs) may hide actual costs from consumers. The *net cost* of a good is the combination of actual costs and subsidized costs. Further, it is important to compare the net cost of alternative fuels to the net cost of the oil presently used to produce gasoline and diesel fuels.

What are some of these costs? Let's start with a baseline figure: The United States currently spends about $400 billion per year to import oil, up from about $300 billion in 2004. This figure does not include hidden costs and social costs.

Let's explore those related costs by turning to the testimony of Mr. Milton R. Copulos before the Senate Foreign Relations Committee on March 30, 2006. Copulos, head of the National Defense Council Foundation (NDCF), testified on "The Hidden Cost of Our Oil Dependence." He similarly testified before the House Resources Subcommittee on Energy and Mineral Resources in March 2004. On

> Subsidies to producers of ethanol and biodiesel are great buys for the U.S. public.

each occasion, his testimony updated the NDCF's detailed 2003 analysis of the total economic cost of the nation's growing dependence on imported oil. The original report from 2003 involved the evaluation of "hundreds of thousands of documents" over more than 18 months. The findings were "vigorously peer reviewed."

An excerpt from his testimony:

> In 2006, we're going to spend about $320 billion to buy imported oil. That's 3.2 times what we were spending three years ago. We feel that the average refiner price will be about $60 a barrel, not $28 and some change. [Note: The cost rose to $99 per barrel in November 2007 and to $117 per barrel in April 2008.] And in contrast to the $49 billion we were spending [in 2003] in the Persian Gulf to defend oil supplies, that figure is now $132.7 billion. And when you add everything together and take the economic consequences into account…that $304 billion [total] in 2003 will increase in 2006 to $825.1 billion. That's

almost twice as much as we're going to spend on national defense this year. It adds the equivalent of $8.06 to a gallon of gasoline when we look at the price that was posted yesterday [April 14, 2006]. That means at the pump—if you were paying the full cost—it would be $11.06 per gallon, meaning that it would cost you about $220 to fill up a sedan and about $325 to fill up an SUV.

Wake up, everybody. That's a hidden cost of $8.06 per gallon—a cost, including subsidies, that comes right out of your pocket. Mr. Copulos is no flake, and his information sources are widely referenced, as are his analyses. If you want another opinion, see Erica Swisher's article in *Ethanol Today*. She calculated hidden costs of $6.45 per gallon of gasoline. This cost will escalate, of course, as the United States depends more and more on oil imports.

For argument's sake, let's use only the lower figure of $6.45 per gallon. If so, then a $1–2 subsidy for ethanol or biodiesel is one super buy.

The U.S. public should not let the ethanol and biodiesel industries struggle for lack of financial support. Let's support them as necessary to keep them vital. As costs fall (and as the costs of alternatives rise), the support should shrink. I recommend paying a temporary surcharge on all energy to support the transition to clean, renewable energy sources.

If the world really wants to avoid the serious consequences some predict will result from the continued use of fossil fuels—global warming, the devastation of ocean pH change, acid rain, mercury pollution, and resource depletion—start the transition to renewables. Yes, doing the right things will displace some people and some industries, but they will also spark an unprecedented economic boom. As we shall see, getting the money is really no problem. Clearing the regulatory, legal, and political barriers are the most formidable challenges. Time is short—30 years and no more.

## BOTTOM LINE

▷ Governments and the public should aggressively support the production of corn ethanol, cellulosic ethanol, soybean biodiesel, and algae biodiesel.

▷ Governments should facilitate faster growth of these industries, particularly cellulosic ethanol and algae biodiesel. Plants should be built immediately—with excessive costs covered by the recommended surcharges.

▷ Governments should do everything they can to support these industries. Good energy policy. Good economic policy.

▷ Quantify the need for biofuels and calculate the effect on food supplies. Such calculations will become more important with population growth and as production of these fuels accelerates.

*Chapter 9*

# OTHER RENEWABLE ENERGY SOURCES

In 2002 biomass and other renewable sources of energy provided approximately 3.5 percent of total U.S. energy. This figure rose to about 5.5 percent by 2006. Of this total, biomass and hydro-electric power generated 92 percent, geothermal 5 percent, wind 2 percent, and solar 1 percent. The production of energy from renewable sources is growing, but has a long way to go to significantly affect growing U.S. energy needs.

## DAMS and HYDRO-ELECTRIC POWER

A dam is the liquid counterpart to a windmill. One relies on the movement of air; the other relies on the flow of water. Wind power and hydro-electric power are relatively clean and close to being eternal, because nature continuously moves air and water.

There are approximately 75,000 dams in the United States affecting approximately 600,000 miles of river, about 17 percent of the total river length in the United States. Hydropower accounts for approximately

2.7 percent of U.S. electric energy. Many dams no longer make economic sense and should be decommissioned.

In the developed world most good sites for large hydro-electric power plants have already been exploited. However, many dam projects, some quite large, are being built in the developing world. The most notable example is the huge project on the Yangtze River in China. In the process of building a dam, large areas of land are put under water to create the reservoir required to hold the water to be released in a controlled way to generate electricity. The people relocated in China for the Three Gorges project had lived on that land for centuries before it was flooded—a culture shock for the people involved.

Hydro-electric power does not require a dam. A water turbine placed in a fast moving stream can also produce electricity.

A great advantage of hydropower is that it doesn't generate any carbon dioxide, carbon monoxide, sulfur dioxide, nitrous oxide, ground contamination, particulates, or waste products. The energy is obviously renewable, since the dam is replenished by rain, snow, and runoff. Of course, if the river dries up due to drought or climate change, then you are out of business until the water to operate it is replenished.

The output of hydro-electric power plants can be controlled at will, and the reservoir lakes can be used for recreation. The reservoir lakes created by the Tennessee Valley Authority and at Lake Powell in Utah, created by the Glen Canyon Dam, are good examples.

As with most energy sources, hydro-electric power has some negative features. I've already mentioned the displacement of people and the problems posed by drought. In addition, should a dam break by natural causes or sabotage, the rapid, catastrophic downstream flood would cause considerable damage and loss of life. Of course, dams are built with structural safety factors. Another problem has to do with the ecosystem around large dams and sedimentary buildup. Since dams impede river flow, low oxygen levels in the reservoirs can kill fish and affect the nature of nearby plant and animal life. Spawning fish can also be a problem.

All in all, once a dam and a new surrounding ecosystem get established, a dam is a wonderful source of renewable energy. All dams, however, have finite life spans—100–200 years for some, perhaps 1000 years for others, such as Hoover Dam.

## TIDES

Much has been written about other forms of hydropower. One form takes advantage of the difference in sea level between high and low tides, and another type uses the motion of waves to generate electricity. I encourage taking advantage of both energy-generating systems, but only in light of the full life-cycle costs of such projects. While these systems promise minimal environmental impact, the world must be wary of unintended consequences.

Tides are reliable and renewable. There are many places with large differences between high and low tide, where flow can be managed to give consis tent energy. Tides, as dams and waves, produce no polluting gases or other wastes. Tidal electric-power systems are simple, and the turbines are expected to last more than 30 years. While I cannot find reliable quantification, the initial cost of a system is high. One cost problem is that the best tidal sites are often the most treacherous. Another problem is fish and other marine life causing damage to the turbines or to themselves. Other problems, such as sediment or modification of the clarity of the water, can be significant. The total effects on an ecosystem depend on the specific site and location. In summary, tidal energy makes sense in some locations.

Tide-turbine electricity costs about the same as wind-turbine electricity. Authorities are planning a total of 100–300 turbines for Roosevelt Island in the East River of New York City. The 3-bladed turbines, each 16 feet in diameter, are made by Verdant Power of Arlington, Virginia. According to the MIT *Technology Review* (April 23, 2007), at full capacity the project could produce 10 megawatts of electricity, enough for 8000 homes in New York.

## WAVES

Wave motion and consequent swells can turn a turbine, which then produces electricity. Wave motion is clean, definitely renewable, but variable. Wave-motion systems are not particularly expensive to install or maintain. Also, they blend in well with their surroundings because of their low profile, and they do not unreasonably or adversely affect marine life. Large systems can produce great quantities of usable electricity.

Like wind, wave motion is not constant. Therefore, the output is intermittent and must be used in conjunction with some back-up source of power. Also, wave-motion systems can be annoyingly noisy and can, because of their low profile, present a hazard to navigation. Many analysts speculate that to make these systems safe from a once-in-a-lifetime storm, they would become prohibitively costly. As the world moves forward with these systems, technological innovations should, over time, reduce the cost and mitigate some problems.

Again, while these hydro-electric sources of energy will help, they cannot rescue us—far from it. However, I hope that the world continues to find places and ways to exploit all sources of hydropower.

## GEOTHERMAL ENERGY

"Geothermal" in Greek literally means "earth heat." The earth's core is extremely hot as a result of radioactive activity deep in the earth. Wherever you go, you will find heat at a reasonable depth. In some places this heat is exploited as a source of energy. The temperature of the earth increases about 80 degrees Fahrenheit for every mile of depth. If you drill deep enough you find magma, which is molten rock. Volcanic eruptions bring this magma to the surface. However, if one wanted to build a geothermal energy plant, one would look for the hottest temperature that is closest to the surface.

Geothermal energy offers many advantages, and I do not know why the United States and others in the world have not done more to exploit this resource. Iceland generates 25 percent of its electrical energy and almost 90 percent of its hot water from geothermal. Obviously the potential of geothermal energy is huge. (See Figure 9.1.) There must also be serious reasons why this energy source is not more aggressively pursued.

## Figure 9.1. Hottest Known Geothermal Regions

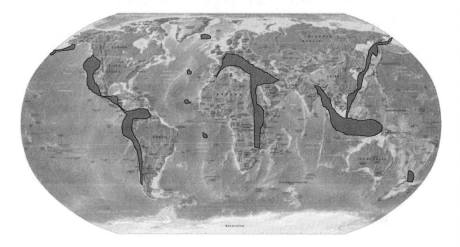

Source: Geothermal Education Office.

Geothermal energy is essentially renewable and eternal because of its long, potential life. Geothermal energy is clean and requires no fuel, because even the energy required by its pumps is derived from the resource itself. Once built, a geothermal plant requires relatively little cost to operate, maintain, or repair. A geothermal plant does not take a lot of space, and when placed next to the ocean, it can desalinate water.

> Geothermal energy seems plentiful, uncomplicated, and likely very economical in places.

As steam is made to drive the generating turbines, the steam must condense. This condensate is saltless and can be used for many purposes, including drinking and irrigation.

Geothermal energy sure sounds like a freebie, and I encourage further development of geothermal sources. However, finding good sites for a plant is a challenge, and getting approval from local residents or governments can be difficult. One wonders if such decisions are made rationally, based on sound science and careful study, or if they are based purely on political considerations, or by misdirected activists.

Some issues relating to geothermal energy must be managed carefully. Subterranean rocks can release toxic and flammable gases which could rise up in the wells, just as gases rise during a volcanic eruption. In some cases natural gas or crude oil will come up in the well, reducing overall costs. In other cases it is difficult to manage this effluent, which can run up operating costs to an unacceptable level.

It seems that geothermal energy should be given a harder look around the world. It seems plentiful, uncomplicated, and likely very economical in places. I suspect that geothermal possibilities have not been sufficiently explored. Low-hanging fruit?

## OTHER

Other energy sources, such as wood and biomass, exist in many forms. I discuss ethanol and biodiesel elsewhere. Other renewable energy sources are not worth mentioning because they are insufficient to make a meaningful difference in the world's quest toward energy independence.

## BOTTOM LINE

- ◖ Economic deployment of the renewable energy systems discussed in this chapter should be part of our overall energy strategy, but none seem capable of making much of a contribution.

- ◖ One must conduct sufficient study to determine if a new renewable source of energy will cause more damage than the energy source it is replacing.

## Chapter 10

# NUCLEAR ENERGY

PROVIDING A CLEAN, AFFORDABLE FUTURE

Only nuclear energy can provide enough clean, reliable energy to accommodate the earth's growing population and development needs. Absolutely no other power source can do it. No other source. No other choice.

## NO OTHER SOURCE:
## THE NUCLEAR ENERGY POWERHOUSE

Only nuclear energy delivered by modern fast neutron reactors can rescue the world from energy disaster—simple as that. Only fast neutron reactors can generate the necessary nuclear energy cleanly, reliably, and affordably. Light-water reactors can't do it.

*Light-water reactors*, the ones mostly in operation today throughout the

> Only nuclear energy delivered by modern fast neutron reactors can rescue the world from energy disaster—simple as that.

world, will not solve our problems because they are unsustainable. They use less than 1 percent of the energy available in the uranium fuel. This wasteful practice of using only 1 percent of the fuel, then sending the rest to a mountain for disposal, will cause the world to run out of uranium in 50–100 years, and to run out of places to store the spent fuel. Thus, the world faces the unavoidable, long-term need for fast neutron reactors and safer recycling.

Nuclear energy from *fast neutron reactors* (also known as integral fast reactors) is essentially eternal and environmentally sound. Nuclear energy from these reactors is *eternal* because it can power the world's needs for more than 100,000 years or, as one scientist put it, until the sun engulfs the earth. Nuclear energy from fast neutron reactors is also *environmentally sound*, because these power plants produce far less of the dangerous nuclear wastes produced by light-water reactors, and the waste is less toxic. The waste ("spent fuel") remains toxic for only 300–500 years rather than for more than 10,000 years. Also, the proposed Global Nuclear Energy Partnership (GNEP), when consummated, guarantees safer, more responsible use of nuclear energy by all partner nations.

The requirements are clear: The United States and the world need a lot of fast neutron reactors. (I offer details below about fast neutron reactors and the Global Nuclear Energy Partnership, and I quantify this point with costs and a timetable in Chapter Sixteen.) There are issues to be sure, as there are issues with all energy sources, but all issues surrounding nuclear energy are manageable. There are also trade-offs among energy sources, but for the next 100 years nuclear energy generated by fast neutron reactors provides the best balance by far. I'm sure of it, as are lots of premier scientists.

Many former opponents of nuclear energy have become wholehearted advocates. Even some environmentalists are embracing nuclear power. Environmental sage James Lovelock sparked a debate in England in 2006 when he published an impassioned defense of nuclear energy—on environmental grounds. "I am a Green," he wrote, "and I entreat my friends in the movement to drop their wrongheaded objection to nuclear energy." His argument: Splitting atoms is the only way to generate huge quantities of electricity without producing the volumes of global-warming gases emitted by plants fired by coal or natural gas.

Even in the United States nuclear power is growing in respectability. According to Nicholas Varchaver in *Fortune* magazine, "the bipartisan National Commission on Energy Policy included it [nuclear power] in a December [2004] proposal to 'end the energy stalemate.'" Columbia University's Earth Institute considers nuclear energy an option in its State of the Planet assessment. And Richard Smalley, a Nobel Prize-winning chemist at Rice University who has been delving into energy issues, echoes Lovelock: "We ought to, and probably will, start building nuclear power plants again."

The United States and the world better get started. We don't have much time. We will stand at the edge of an imminent disaster if we don't get moving. Given the rapid depletion of fossil fuels and the rapid accumulations of pollutants, it will be too late for a smooth transition to alternative energy sources if we don't immediately launch a massive, national and international nuclear-energy program. Any plan short of immediate, aggressive action will put the economies of the world in a tailspin.

> Any "debate" about whether nuclear power will be used to generate a major portion of the world's electricity is really over.

The world still needs and must continue to build light-water reactors until fast neutron reactors are ready, which I hope will be very soon. Russian submarines are powered by fast neutron reactors. Electricity-producing, fast-neutron-reactor plants include the Superphénix, the Fermi, and the Monju; however these are not of the most recently proposed design.

Scientists have demonstrated the efficacy of the process that couples an advanced fast neutron reactor with the appropriate recycling of the spent fuel—fuel otherwise destined for Yucca Mountain, Nevada. Scientists in several countries, including the United States, want to pilot the entire process before proceeding to final, optimum plant design. Others don't believe a pilot plant is necessary; they think we can move directly to commercial plant design. Why argue? Let's build the pilot plant now so the very stuff you worry about can be used to fuel fast neutron reactors rather than fill Yucca Mountain.

The combination of fast neutron reactors and the Global Nuclear Energy Partnership—in one fell swoop—minimizes today's problems of spent-fuel storage and proliferation.

Any "debate" about whether nuclear power will be used to generate a major portion of the world's electricity is really over. Distracting nuisances who vigorously criticize nuclear energy—notably the Union of Concerned Scientists and Helen Caldicott—are too late. There are already 441 reactors operating in the world—and more on the way—and 103 operating in the United States. The storage of spent fuel is manageable. Proliferation is manageable. The combination of fast neutron reactors and the Global Nuclear Energy Partnership—in one fell swoop—minimizes today's problems of spent-fuel storage and proliferation. What are we waiting for? What's left to debate? Nuclear energy from fast neutron reactors takes care of every energy issue facing the world today—depletion of oil, pollution from fossil fuels, and even the prospect of global warming from greenhouse gases.

## NO OTHER CHOICE, NO OTHER OPTION

I acknowledge that solar, wind, and hydrogen energy may contribute more significantly to the world's total energy mix *in the distant future*, but I would not bet much on the prospects. None of the promising energy research that I know of is going to yield results that are timely or sufficiently robust to rescue the world from its *present* energy problems. Nuclear energy must be the backbone of any viable future energy system. Period. I invite your disagreement, but don't bother me unless you quantify your comments and offer a realistic timeline.

Wind and sun can never provide *baseload* energy, because the sun doesn't shine all the time and the wind blows intermittently. Each source needs a back up since, by definition, a baseload source must operate continuously. Wind energy and solar energy, because they are intermittent, can be only

> Nuclear energy from fast neutron reactors resolves every energy issue facing the world today—depletion of oil, pollution from fossil fuels, and even the prospect of global warming from greenhouse gases.

partial substitutes for nuclear, hydro, or fossil-fuel power plants. Solar and wind power can reduce total toxic emissions if they are backed up by fossil-fuel power plants. However, if nuclear plants provide the back-up, then the wind and solar energy are redundant and unnecessary, because nuclear energy is cheaper and cleaner. Wind and solar energy could contribute electricity during high-demand, peak periods—that is, for *peakloads*—and for some site-specific applications.

There's more to the story: A growing consensus holds that wind and sun together will probably not account for more than about 20–40 percent of total energy production, at least not in this century. So, before we get too excited or confident about their potential contributions, let's first see them generate 10 percent of total energy production, a daunting and ambitious task. Although I encourage it, I won't believe the 10 percent until I see it.

In the meantime, be wary of breakthrough announcements. A "breakthrough" is just the first adventurous step on the journey to commercialization. Such journeys can take decades.

But none of nuclear energy's virtues really matter if nuclear energy is too dangerous to deploy. *Is nuclear energy safe?*

❯ Yes, nuclear energy has been the safest energy source by far over the last 50 years.

◖ Yes, nuclear power plants are absolutely safer than previous generations because of new technologies, new reactor designs, and better process management.

◖ Yes, because fast neutron reactors will be safer than reactors built in the past.

◖ Yes, since nuclear energy is definitely safer than the continued use of fossil fuels, which annually kills 2 million people worldwide, 50 thousand in the United States, and gets worse every year.

◖ Yes, because the spent fuel problem essentially goes away with fast neutron reactors.

◖ Yes, because the proposed Global Nuclear Energy Partnership will establish a fuel- and waste-management process that will greatly decrease proliferation opportunities and the misuse of nuclear materials.

## NUCLEAR ENERGY BACKGROUND

Many of us grew up fearing nuclear energy: the secretly developed bombs that put an abrupt end to World War II, the Three Mile Island accident in the United States, the Chernobyl accident in Russia, the threat of annihilation from nuclear weapons, and proliferation fears. Most of us, however, do not know much about the real outcomes of these accidents, about nuclear power's ability to generate electricity, or about the many other useful applications of nuclear technology.

Except for research on nuclear energy, none of the promising energy research that I know of is going to yield results that are timely or sufficiently robust to rescue the world from its present energy problems.

Despite the public's lack of familiarity, a large majority of the public supports nuclear energy. As long ago as February 1989, a Gallup poll asked "How important do you think nuclear energy plants will be in providing this nation's electricity needs in the years ahead?" The poll also asked exactly the same question about coal-burning plants. Forty-five percent of respondents thought nuclear energy would be "very important" (29 percent for coal), and 34 percent thought nuclear energy would be "somewhat important" (37 percent for coal). That is, 79 percent of respondents thought nuclear energy would be important. A similar poll in July 1989, from TeleNation Market Facts, asked, "How important a role should nuclear energy play in the U.S. Department of Energy's National Energy Strategy for the future?" The results were virtually identical. Among respondents, 50 percent thought nuclear energy "important," and 31 percent thought it "somewhat important."

> Scientists and experts are very strong supporters of nuclear energy.

Although one might guess that fears of nuclear energy have diminished as the memories of Three Mile Island and Chernobyl fade, public support in the United States today is lower than in the 1980s, although supporters are still a clear majority. Various polls conducted in the last few years report 50–70 percent of the U.S. population supports nuclear energy.

More importantly, the people who know more about this technology—scientists and experts—are very strong supporters of nuclear energy. In 1980, in the immediate aftermath of the Three Mile Island accident and during a strong wave of anti-nuclear sentiment, a random sample of scientists listed in *American Men and Women of Science*, the "Who's Who" of scientists, received 741 responses to questions about nuclear energy. One question asked, "How should we [the nation] proceed with nuclear power development?" Responses appear in Figure 10.1.

> Nuclear power is the only energy source that can fully supply all the clean energy the world will ever need.

## Figure 10.1. Percentage Support for Nuclear Energy Among Scientists, 1980

| | All Scientists | Energy Experts | Nuclear Experts |
|---|---|---|---|
| **Proceed Rapidly** | 54 | 70 | 92 |
| **Proceed Slowly** | 36 | 25 | 8 |
| **Halt Development** | 7 | 4 | 0 |
| **Dismantle Plants** | 3 | 1 | 0 |

With such support for nuclear energy among experts and the general public, why so little progress on this elegant form of energy? Politicians and decision makers can't make it happen. Or they don't want to make it happen. You decide. Are vested interests trumping popular support and the public good? This strong support existed even before the development of safer modern fast neutron reactors, nuclear batteries, or the proliferation-resistant UREX+ processing. Nuclear energy should be even more attractive now than in 1980 and 1989.

## HOW ELEGANT IS NUCLEAR ENERGY?

One pound of uranium converted to energy produces 3.8 million times more energy than a pound of coal. One golf-ball-size chunk of uranium fuel converted to energy can provide 4 people with a lifetime's worth of energy for electricity, transportation, and heating. For a breathtaking comparison, the same job requires 8000 barrels of oil, *and* 22 million cubic feet of natural gas, *and* 1200 tons of coal.

The process of nuclear fission is awesome, beautiful, elemental, and elegant. Albert Einstein changed the world when he quantified the energy released when converting mass to energy. This conversion is defined by the formula $E = MC^2$. As a simple translation, certain isotopes of some elements lose weight in a nuclear reactor. The lost weight represents the material that no longer exists because it was converted to energy. Einstein's genius assures sufficient energy for future generations. We can't argue with the basic chemistry or physics or with the realities and limitations of all other proposed energy sources: Nuclear is the only energy source that can fully supply all the clean energy the world will ever need.

## HISTORY of NUCLEAR ENERGY in the UNITED STATES
### Origins

On December 2, 1942, Enrico Fermi and Leo Szilard, both at the University of Chicago, first built a nuclear pile and demonstrated a controlled nuclear chain reaction. Not long afterward the United States and the Soviet Union were making nuclear bombs, long before nuclear reactors were generating electricity. It is well known that a country does not really need a reactor to make uranium-based atomic bombs. In the mid-1950s, the Soviet Union and Western governments secretly expanded nuclear research to include non-military uses of the atom. On December 20, 1951, electric power from a nuclear-powered generator was produced for the first time, near Arco, Idaho.

*The 103 operating reactors in the United States generate 22 percent of U.S. electricity and about 80 percent of U.S. carbon-free electricity.*

The U.S. Navy—under the leadership of one of my heroes, Admiral Hyman Rickover—became the first organization to develop useful nuclear power for the propulsion of submarines and, eventually, of aircraft carriers. On January 17, 1955, with the words "Underway on nuclear power," Commanding Officer Eugene Wilkinson ushered in the U.S. Navy's nuclear era by reporting the launching of the USS *Nautilus*, which continued in service until 1980. The U.S. Navy has operated more nuclear reactors than any other entity in the world, with perhaps the exception of the Soviet Navy. These reactors have operated with no—zero—serious mishaps. Two submarines, the USS *Scorpion* and USS *Thresher*, were lost at sea with no expectation of harmful environmental effects. *Nuclear power permits submarines and carriers to be at sea for close to 20 years without refueling. Newer ships may operate for over 30 years without refueling.*

## Naïve Nuclear Policy in the United States

In the United States the most significant problem with developing and using nuclear energy is a big one—a political problem. Not an engineering problem. Not a chemistry or physics problem. A political problem! The United States, once the world's leader in nuclear technology, more or less went out of the nuclear plant business after building over 100 light-water reactors. To a large extent the United States has since abandoned the ongoing development of nuclear technology. Our leaders lacked vision and resolve, and protestors caused project-killing delays. These delays amounted to a vote *for* dirty coal and a fouler environment.

The decision of the Carter Administration to end the reprocessing of spent fuel ably illustrates political failings. The choice was based on the naïve belief that if the United States didn't reprocess, then the rest of the world wouldn't. PUREX reprocessing (Plutonium and Uranium Recovery by EXtraction), the reprocessing method available when Carter issued his ban, can increase the use of the original *uranium* fuel by about 20 percent. However, PUREX reprocessing can also produce *Plutonium-239* of sufficient chemical purity to make a weapon. Sufficient concentrations of Plutonium-239, necessary to construct a plutonium-based nuclear weapon, are obtainable only via reprocessing spent fuel. However, a ban on reprocessing does not end the opportunity to create nuclear weapons; one can devise a uranium-based nuclear weapon by "enriching" uranium.

A second problem was the success of U.S. anti-nuclear activists in killing the nuclear industry by causing construction delays. These delays created astronomical cost overruns and often caused the abandonment of projects and of the billions of dollars invested. Even before the 1979 accident at Three Mile Island, new orders for nuclear plants in the United States had ceased, primarily because of construction delays caused by court-mandated injunctions. Some plants were never completed. As of 2007, no new nuclear plants have been started in the United States for almost 30 years—an unbelievable lack of foresight. I believe a terrible flaw in our legal system permits a small minority to pre-empt what the majority wants and needs.

## Present Position

The 103 operating nuclear power plants in the United States are located in 31 states. These plants generate 22 percent of America's electrical energy, all essentially emission-free. This is more electricity than the total electrical energy consumed by France and Spain together. To generate the same amount of energy from coal, you would have to mine, transport, and burn about 400 million tons of coal. If a coal car is 90 feet long and carries 60 tons, then a train to transport 400 million tons of coal would be 110,000 miles long.

One golf-ball-sized chunk of uranium fuel does the same job as 8,000 barrels of oil, *and* 22 million cubic feet of natural gas, *and* 1200 tons of coal.

Governments and the public throughout the world must support the nuclear-energy industry. What should the United States do? The United States must vigorously add conventional light-water reactors to our reactor fleet *now*. Within 5–7 years we must be building commercial fast neutron reactors. These reactors are presently scheduled to be ready in 20 years. Not good enough. The United States needs these reactors now. General Electric, Toshiba, and some other companies already have modern fast-neutron-reactor designs ready to go— there would be no delay between pilot-plant operation and commercial construction. Also the world must launch the Global Nuclear Energy Partnership as soon as possible.

Most specifically, the United States must build the fast-neutron-reactor and recycling pilot plants immediately, now, today. The Russians will likely build their own pilot plant, and the United States may partner with them. At best, this plan promises red tape, delays, and compromise. Instead, the U.S. government should immediately build a pilot plant using technology already developed in U.S. government labs. The cost would be less than $6 billion, which could be funded from energy surcharges I recommend (see Chapter Sixteen). In the interest of time and technology, I think U.S. National Laboratories— the labs that did the development work—should run the pilot program. At a minimum the National Labs should partner with a commercial company. Such an investment would save trillions of dollars over time; delays could easily cost the country additional tens of billions of dollars per year.

> The United States cannot remain a world power—or provide a reasonable standard of living for our children and grandchildren—without abundant, safe, inexpensive energy.

We need no more studies, which are often window dressing, an excuse to do nothing. We need action. Government must do everything it can to accelerate these programs. *In all energy matters, we must bring the future closer to the present as quickly as possible.*

It takes only money and commitment. We have the money. Commitment has been much harder to come by, particularly in Washington, D.C.

In his speech "Reminiscences of Reactor Development at Argonne National Laboratory," Charles E. Till, Associate Laboratory Director, 1980–1998, reports that the Clinton Administration stopped the advanced fast-neutron-reactor pilot program. In my view, this was an unforgivable mistake, for which Americans will pay dearly. Till claims the Administration made its decision due to a lack of understanding and political motivations, rather than technical concerns. In fact, the technical matters were humming along with no problems. I quote Till at length.

> Two elements of the system—the fuel and the fuel cycle—were our principal focus. They were the necessary first steps. But by 1986, we had also prepared for many months for a series of demonstrations of the unusual safety characteristic made possible by the excellent heat-

transfer characteristics of metallic fuel and the liquid sodium coolant. It made unaided shutdown of a properly designed reactor under accident conditions possible—power reduction and shutdown just from the interplay of the heat transfer characteristics of the new fuel with the sodium coolant. No operator action, no operation of safety systems, would be needed just to ride through the two major accident-initiating events: *Loss of Heat Sink*, as in the [Three Mile Island] accident, and *Loss of Flow*, an accident possibility that at the time had not occurred in any power reactor, but which had long been studied.

In early April of that year, both accident cases were initiated in our test reactor EBR-2—both while at full power. In the morning, the reactor was suddenly isolated from the steam system, cutting off the heat sink. The reactor responded by smoothly shutting down. Then in the afternoon, after starting up again, the pumps were turned off; the flow coasted down, but

> We have the money. Commitment has been much harder to come by, particularly in Washington, D.C.

… so did the power—in lockstep with the flow coastdown. *In both tests the reactor had quietly shut itself down* [emphasis added]. DOE duly issued a press release.

Nobody paid any attention.

Then the loss of flow accident happened. And it happened on the world stage, with riveting TV coverage, and the greatest possible concern—at Chernobyl.

An alert science reporter at the *Wall Street Journal*, Jerry Bishop, made the connection immediately. He remembered the press release and he made the connection himself. A reactor in Idaho had lost its coolant flow, and at full power, in this same month, and NOTHING WHATEVER had happened. He contrasted this with the tragedy unfolding at Chernobyl.

The Bush Administration revived the fast-neutron-reactor program, but the 2006 Democratic Congress blocked funding. The Energy Policy Act of 2005 proposed funds for up to 6 new nuclear reactors. This was too few reactors, and funding issues remain.

We need no more studies, which are often window dressing, an excuse to do nothing. We need action.

Republicans do no better. According to the *Minneapolis Star Tribune* (July 12, 2007), Norm Coleman, Minnesota's Republican senator for the 2002–2008 term, in an effort to maintain his public visibility, played up the possibility of a dirty nuclear bomb in another cheap, sensational, unwarranted nuclear scare.

We need a president and Congressional representatives who are informed and able to address the national and global energy problems. Candidates preening for the 2008 national elections just don't seem to "get it." I fear they will continue fumbling the most important issue of this century. Any candidate from any party who is not prepared to deal with our energy problem is not worthy of public office.

U.S. leaders spend untold billions and send young men and women to Iraq to fight and die for freedom and oil, which won't solve U.S. energy problems anyway. But these same leaders are unwilling to spend a few billion dollars to assure the country's future well being. Build the pilot plant. The United States cannot remain a world power—or provide a reasonable standard of living for our children and grandchildren—without abundant, safe, inexpensive energy. Business-as-usual politics will not get the job done.

Other nations are on the move, and what the United States does is becoming less and less relevant.

## NUCLEAR ENERGY WORLDWIDE

According to the World Nuclear Association, the world has 441 operating reactors, 33 reactors under construction, 94 reactors planned for future construction (most with improvements upon traditional designs), and 222 proposed reactors. China, the most active by far, has 5 reactors under construction, 30 planned, and 86 proposed. The United States has zero under construction, 7 planned, and 25 proposed, but it remains a political question whether the United States will ever build these plants. Nevertheless, the United States has more operating reactors than any other country (103 of the global total of 441).

Over 20 percent of U.S. electrical energy and 17 percent of the world's energy comes from nuclear-power plants. France generates 78 percent of its electrical energy from 59 nuclear plants, and Lithuania generates 72 percent of its electricity from nuclear plants. Five countries generate over 50 percent of their electrical energy from nuclear sources, and 8 countries generate 30–50 percent. Japan, currently generating 29 percent of its national electricity from nuclear plants, plans to go to 35 percent; the United States, now at 22 percent, plans to go to 23 percent.

Other countries plan big expansion programs. By 2050 China plans to expand its production of nuclear-generated electricity by 240 gigawatts, which is almost equal to China's total electrical consumption in 2004. India's plans are similar. Such production represents approximately 2.5 times all U.S. nuclear-produced electricity today. Japan, China, and India are all actively developing the technologies for fast neutron reactors and are building light-water (thermal) reactors. South Korea, South Africa, Finland, France, and others are developing advanced nuclear technologies. Other nations are on the move, and what the United States does is becoming less and less relevant. However the entire world is not building nuclear plants anywhere near fast enough.

In 2003 the United States spent $49 million on nuclear fission research. France spent $394 million. Japan spent $2.3 billion. The

United States has seemingly opted out of the most important energy technology for the world, although its fast neutron reactor technology is still among the world's best. The United States must not permanently opt out. It must regain its leadership role.

Since nuclear energy is here to stay, all nations should help each other manage it safely. A significant concern is the possible proliferation of nuclear *weapons*. However, the problem is not with electrical-energy-producing reactors. Rather, the problem is all the bombs and highly enriched uranium and plutonium lying around the world. The good news is that new nuclear plants, particularly fast neutron reactors, are better, more efficient, more proliferation-resistant, and safer than the plants already in service. So let's spend our collective time and effort on making nuclear *energy* even safer as we accept the inevitability of fast neutron reactors—as we must. The adoption of advanced fast neutron reactors and of the proposed Global Nuclear Energy Partnership would be giant steps toward solving the world's most daunting challenge—energy independence for all nations—and freedom from the dire consequences of burning fossil fuels.

Why are we still arguing over whether nuclear energy has a future?

> Any candidate from any party who is not prepared to deal with our energy problem is not worthy of public office.

## NUCLEAR ENERGY and the MEDIA

Concerns with nuclear energy have been relentlessly pounded into the public by the media and some of the groups they quote, such as the Union of Concerned Scientists. I don't blame the media for quoting the rants of Chicken Littles about the-sky-is-falling and nuclear-energy-will-doom-us-with-radiation. Those sensational claims sell. Rather, the media could do a great service to the country and the world by presenting credentialed points of view by real experts, particularly on complex, science-based topics that the writers themselves often do not understand. It is very difficult to debate these very complex technologies in the popular media driven by sound bites and catchy headlines to appeal to a public with a short attention span and a limited appetite for complexity.

The media should call for and provide real debate by real experts on issues that could seriously affect the future of the country. Some of the best, real, credentialed, scientific experts I interviewed for this book felt they have *no voice* in the media, *no voice* to the public, and *no voice* in making policy. On top of that, we have know-it-all leaders who are technically illiterate. The world seems impressed by Al Gore because he has a better-than-basic understanding of some issues. Yet he's wrong on some key points, and his solutions and proposals are anemic, if not ridiculous. Imagine what other politicians know or think they know. This is a horribly sad state of affairs in a technology-driven world.

Sometimes very vocal groups with strong opinions based on inaccuracies and too few facts change their minds, but often only after causing considerable damage. Dr. Patrick Moore, a co-founder of Greenpeace, said, "Rather than promote unilateral boycotts based on misinformation and coercion, organizations like Greenpeace should recognize the need for internationally accepted criteria." Moore also said, "Fifteen years of Greenpeace later, I had some new insights. It was time to switch from confrontation to consensus, time to stop fighting and start talking with the people in charge."

> "Nuclear energy is the only non-greenhouse gas emitting power source that can effectively replace fossil fuels and satisfy global demand."

Dr. Moore also made the following landmark statement on April 28, 2005: "Nuclear energy is the only non-greenhouse gas emitting power source that can effectively replace fossil fuels and satisfy global demand." How many of you read this comment in the news? Did you even hear about it? Of course not.

## Excerpts from Dr. Moore's Testimony to a Congressional Subcommittee on Nuclear Energy, April 28, 2005

What does environmental extremism have to do with nuclear energy?

I believe the majority of environmental activists, including those at Greenpeace, have now become so blinded by their extremist policies that they fail to consider the enormous and obvious benefits of harnessing nuclear power to meet and secure America's growing energy needs.

These benefits far outweigh the risks.

There is now a great deal of scientific evidence showing nuclear power to be an environmentally sound and safe choice.

Today nuclear energy supplies 20 percent of U.S. electrical energy.

Yet demand for electricity continues to rise and in the coming decades may increase by some 50 percent over current levels.

If nothing is done to revitalize the U.S. nuclear industry, the industry's contribution to meeting U.S. energy demands could drop from 20 percent to 9 percent.

What sources of energy would make-up the difference?

It is virtually certain that the only technically feasible path is an even greater reliance on fossil fuels.

Dr. Moore goes on:

In a "business as usual" scenario a significant reduction in greenhouse gas emissions (GHG) seems unlikely given our continued heavy reliance on fossil fuels. An investment in nuclear energy would go a long way to reducing this reliance and could actually result in reduced carbon dioxide emissions from power generation.

According to the Clean Air Council, annual power plant emissions are responsible for 36% of carbon dioxide ($CO_2$), 64% of sulfur dioxide ($SO_2$), 26% of nitrogen oxides (NOx), and 33% of mercury emissions (Hg).

These four pollutants cause significant environmental impact, including acid rain, smog, respiratory illness, mercury contamination, and are the major contributors to GHG emissions.

Among power plants, old coal-fired plants produce the majority of these pollutants. By contrast, nuclear power plants produce an insignificant quantity of these pollutants.

**"Technology has now progressed to the point where the activist fear-mongering about the safety of nuclear energy bears no semblance to reality."**

According to the Clean Air Council, while 58% of power plant boilers in operation in the U.S. are fueled by coal, they contribute 93% of NOx, 96% of $SO_2$, 88% of $CO_2$, and 99% of the mercury emitted by the entire power industry.

*Prominent environmentalists see nuclear energy as a solution. Prominent environmental figures like Stewart Brand, founder of the* Whole Earth Catalog, *Gaia theorist James Lovelock, and Hugh Montefiore, former Friends of the Earth leader, have now all stated their strong support for nuclear energy as a practical means of reducing greenhouse gas emissions while meeting the world's increasing energy demands.*

I too place myself squarely in that category.

U.K. environmentalist James Lovelock, who posited the Gaia theory that the Earth operates as a giant, self-regulating super-organism, now sees nuclear energy as key to our planet's future health. [ Lovelock, the author of over 200 scientific papers, has been described "as one of the great thinkers of our time" (in *New Scientist*) and as one of the world's top 100 global public intellectuals (by *Prospect*).] "Civilization is in imminent danger," he warns, "and has to use nuclear—the one safe, available energy source—or suffer the pain soon to be inflicted by our outraged planet."

As Stewart Brand and other forward-thinking environmentalists and scientists have made clear, technology has now progressed to the point where the activist fear-mongering about the safety of nuclear energy bears no semblance to reality.

The Chernobyl and Three Mile Island reactors, often raised as examples of nuclear catastrophe by activists, were very different from today's rigorously safe nuclear energy technology.

*Today, approximately one-third of the cost of nuclear reactor is dedicated to safety systems* and infrastructure. ...

I want to conclude by emphasizing that nuclear energy—combined with the use of other alternative energy sources like wind, geothermal and hydro—remains the only practical, safe and environmentally-friendly means of reducing greenhouse gase emissions and addressing energy security [emphases added].

In contrast to Dr. Moore's straight talk, other groups seem more satisfied to convey misinformation and misleading views. I believe one group and one author routinely and seriously mislead the public and have caused much economic and environmental damage by their writings and declarations. I think they violate the public trust with reckless misstatements on a vital subject. I don't begrudge their prejudices, but I do resent misinformation to support those prejudices.

## How I Learned to Distrust the Union of Concerned Scientists

The Union of Concerned Scientists (UCS), a particularly strong and biased critic of nuclear energy, produces often misleading and inaccurate statements to support its scientific and political bias. I am disturbed that the media often preferentially quote the UCS press-release machine that has a dedicated staff belching out information on a wide array of subjects.

The UCS purports to represent the scientific community as a whole and, in their name, to take positions on various scientific and other issues. The group's name implies that the organization represents a large group of scientists. Nothing could be further from the truth. In a dated study, Rothman and Lichter in 1982 found that less than 2 percent of UCS's members are scientists. Two percent is about the same as the percentage of scientists in the general population. Even the group's name is misleading. UCS officials declined to provide information necessary for others to poll the Union's members. The group isn't sure how many members it has. In one place it claims 60 thousand members, in another place 100 thousand members. Sounds like the group needs a recount. Apparently the only requirement for membership is the willingness to submit the membership fee.

If the rank and file members are not scientists, then perhaps the group is directed by renowned and active scientists. Nope. The UCS president and chief executive officer (CEO) is Kevin Knobloch, an English major. The UCS's main nuclear expert, the oft-quoted David Lockbaum, passes himself off as a nuclear expert. He has an undergraduate nuclear engineering degree, and he worked in a nuclear plant 10 years ago—hardly cutting-edge credentials. I was the CEO of

Teltech, a national technical information service that provides technical experts to industry, and I can assure you his credentials would not qualify him as an expert, and certainly not an in-the-know, cutting-edge expert.

The public has unwittingly given some people tacit authority to tell us what we should do and think, but with no responsibility for the outcome if they err or misinform. Authority without responsibility is always dangerous. I think we must put our trust in the country's real experts, those who spend their careers studying all aspects of a technical subject. If you don't trust them, then do you really think it is better to put your trust in those that have a vested interest in business as usual or who are self-proclaimed experts with insufficient background in the science?

## Helen Caldicott

Dr. Helen Caldicott recently published *Nuclear Power Is Not the Answer*. I've not seen a published review of the book, but I cannot imagine any scientist knowledgeable about nuclear power could ignore the mathematical and factual faults throughout. She discounts as wrong and misguided her many high-profile colleagues: Patrick Moore, a former Greenpeace leader; Stewart Brand, founder of the Whole Earth Catalog; and Gus Speth, Dean of Yale's School of Environmental Studies. Caldicott condemns these leaders because they changed their minds and now support nuclear energy. Her disdain strikes me as arrogant, inflammatory, and insulting.

Caldicott makes bold statements with no factual data to support many of them. For example, her Chapter Six expresses her lack of understanding of the recent generation of nuclear reactors. On page 125 and elsewhere she fails completely to comprehend the importance of isotopic quality to plutonium bomb designers. In one glaring misconception, she says only 10 percent of the plutonium is converted into fission products by a fast neutron reactor, yet the correct number is close to 100 percent. (Her 10 percent figure assumes no recycling of fast-reactor fuel, which would defeat the whole purpose of fast neutron reactors.) Pages 126, 127, 128, and 129 are so wrought with errors that a rebuttal would be as long as the chapter. The problems on these pages were identified by legitimate scientists and experts in nuclear energy, not by me.

She also references herself about 20 times, thereby perpetuating her past inaccuracies. In addition she references the Union of Concerned Scientists approximately 40 times and specifically cites David Lockbaum about half of those times.

## THE GLOBAL NUCLEAR ENERGY PARTNERSHIP

The Global Nuclear Energy Partnership (GNEP) is the idea of the century in my opinion. GNEP, the brainchild of the U.S. Department of Energy, offers the most sensible means ever proposed for safely managing nuclear energy worldwide. GNEP promises almost limitless, environmentally friendly energy to all countries around the world that want to deploy nuclear energy for peaceful purposes. The Partnership's ultimate promise is that every country will have adequate energy to grow and prosper. To this end GNEP proposes a global partnership to oversee all aspects of nuclear energy, including the management and control of nuclear fuel and waste. GNEP eliminates or greatly reduces most potentially harmful aspects of nuclear energy and virtually eliminates the risk of nuclear-weapon proliferation. President George W. Bush supported the proposal in his 2006 State of the Union Address.

### What is GNEP?

As detailed by President George W. Bush's Secretary of Energy, Samuel W. Bodman, GNEP has four main goals.

- ❯ Reduce America's dependence on foreign sources of fossil fuels and encourage economic growth.
- ❯ Recycle nuclear fuel using new proliferation-resistant

technologies to recover almost all of the energy in the fuel and to reduce waste.

▶ Encourage prosperity, economic growth, and the development of clean energy around the world.

▶ Use the latest technologies to reduce the risk of nuclear proliferation worldwide.

If the Partnership is fleshed out and carried through to its expected result, then all participating nations will benefit and have a hand in a cooperative effort that will make the world cleaner and safer. Indeed, GNEP will contribute to world peace because nations will have no need to fight energy wars as fossil-fuel resources run out. Within the reach of all nations is energy independence, all the energy they need to grow and prosper in their own way and at their own pace. This proposal constitutes the most profound win-win situation ever proposed on such a large, worldwide scale.

> GNEP will contribute to world peace because nations will have no need to fight energy wars as fossil-fuel resources run out.

The U.S. Department of Energy will work with the Department of State to engage international partners to participate in the GNEP initiative. Over 100 countries have been briefed on the GNEP proposal. The response has been positive, as it should be, because it paves the way for sensible and safe use of nuclear power worldwide. For example, Japan, the country that has suffered the most from the destructive effects of nuclear energy, issued the following statement on February 7, 2006, through its Ministry of Foreign Affairs, on the U.S. proposal for GNEP:

> The Government of Japan welcomes the United States' new initiative to enhance the worldwide development and expansion of nuclear power generation while ensuring nuclear non-proliferation. It is particularly noteworthy that this initiative indicated clearly the orientation of the

U.S. nuclear policy towards the promotion of spent fuel recycling in order to increase the energy efficiency and reduce the volume of radioactive waste. The Government of Japan will study further this initiative with a view toward identifying the potential areas of cooperation.

As part of President Bush's Advanced Energy Initiative, Secretary Bodman announced a $250 million fiscal year (FY) 2007 request to launch GNEP. Congress subsequently appropriated $80 million, a very naïve, uninformed choice. Don't Congressional representatives know what's at stake? For FY 2008, the president requested $405 million to further promote the GNEP, but the House rejected the request, believing that $120 million should be enough. There is no basis for this arbitrary number, nor for the $243 million the Senate proposed. Silly political games. Even though the United States originally proposed it, some observers believe that the U.S. political system will be the main cause for delay in deploying the Partnership. The U.S. public must fight delays and neglect, and it must compel Congress to quickly allocate enough funding to insure a very ambitious timeline. The return on investment will be enormous.

## How Will GNEP Work?

GNEP's partner nations which will supply nuclear fuel (*supplier-partner nations*) would become the "general partners" in GNEP's operations. The supplier-partners will develop a program to provide nuclear fuel and services to other partner nations, thereby allowing those *recipient-partner nations* to enjoy the benefits of abundant, clean, safe nuclear energy in a cost-effective manner. In exchange, all partners—whether developed or developing nations—will commit to forsake enrichment and reprocessing activities, which they won't need. This arrangement further alleviates proliferation concerns.

Participating recipient countries will enjoy 3 substantial benefits:

❯ They will not incur the costs and suspicions concerning the development of their own nuclear technology.

◖ They will not have to invest in expensive reprocessing or recycling equipment.

◖ They will not need waste-disposal facilities.

Yet GNEP suffers from 2 glaring shortcomings: The proposed funding is insufficient and the implementation timeline of 15–20 years is much too long. Instead, the GNEP timeline should match the expected, rapid, global expansion of nuclear energy *and* the rapid depletion of the world's fossil fuels. No let's-do-it-later mentality will work. We need a man-to-the-moon mentality and comparable effort. In my professional experience, most, if not all, complicated projects take *twice the money and twice the time* originally projected. We must at minimum overwhelm the problems with money and talent. As Carl Sagan put it, "We all live in the same house, earth." Our children's well-being will be testament to our foresight or lack of it.

### GNEP's Contributions

The recycling of nuclear fuel—coordinated under GNEP's authority and made possible by the combination of UREX+ recycling and fast neutron reactors—solves several prominent problems simultaneously.

The problems begin with light-water reactors, the most common type of reactors operating today. Nations should however continue to build light-water reactors to replace coal and natural gas for generating electricity. The use of light-water reactors must be temporary because they are unsustainable. A contemporary light-water reactor uses less than 1 percent of the potential energy in the fuel. A measly 1 percent. If these reactors continue to operate as they do today and if we continue to dispose of the waste as we currently do, then we will run out of affordable uranium fuel in 50–100 years *and* long-lived nuclear waste will continue to pile up. So, light-water reactors are merely a temporary improvement over fossil-fuel-fired plants and a bridge to the next generation of nuclear power.

Two elegant technologies set the foundation for the latest

generation of nuclear power. One technology is a recycling process called UREX+. This proliferation-resistant process greatly reduces nuclear waste and produces fuel for fast neutron reactors. Fast neutron reactors are the second elegant technology. By recycling spent fuel from existing nuclear power plants and reusing the material as fuel for fast neutron reactors, these complementary technologies permit almost complete extraction of the energy contained in the original uranium. This combination of technologies delivers other benefits, too.

### Recycling Reduces Waste and Storage Time

By recycling the existing waste from light-water reactors, *no uranium would have to be mined for more than 100 years*. Recycling also means less total waste, less toxic waste, and shorter storage times.

> Rather than store the waste on-site or in some repository such as Yucca Mountain for 10,000 years or more, fast neutron reactors will re-use nearly all of the waste as fuel—the ultimate recycling process.

The proposed recycling process begins with the waste from existing light-water reactors. Rather than store the waste on-site or in some repository such as Yucca Mountain for 10,000 years or more, fast neutron reactors will re-use nearly all of the waste as fuel—*the ultimate recycling process*. Imagine a 100-pound block of such waste. After proper reprocessing at a UREX+ recycling plant, about 5 pounds goes to storage (for only 300–500 years); the other 95 pounds becomes fuel for fast neutron reactors. Subsequently, the small amount of waste (spent fuel) from the fast neutron reactor will also go to storage, where it must remain for only 300–500 years. Since the waste (spent fuel) from fast neutron reactors is less radioactive than waste from light-water reactors, it can be packed more closely together. As a result, Yucca Mountain can hold 5 times more waste than originally planned.

### Recycling Reduces Proliferation Concerns

The GNEP proposes that reprocessing and recycling occur in only a few, secure supplier-partner countries and only via UREX+. Under some specific conditions, PUREX reprocessing of spent fuel could produce weapons-grade material; therefore, the world is better off if all nations discontinue using this method. The proposed UREX+ recycling process (not the PUREX process) does not produce pure Plutonium-239, the best material for making nuclear weapons.

Scientists at U.S. National Laboratories have successfully demonstrated the proliferation-resistant UREX+ processes. Scientists know the cost of these processes will be competitive with existing technologies, and that the recycled fuel will not be useful to any terrorist or rogue nation wanting to build a bomb. Once the new recycling technologies—UREX+ and pyroprocessing—are successfully piloted in conjunction with a fast neutron reactor to optimize performance, the world will have all the eternal, clean energy it will ever need.

Two elegant technologies set the foundation for the latest generation of nuclear power: UREX+ recycling and fast neutron reactors.

## BOTTOM LINE

- Fast neutron nuclear reactors are inevitable. They represent the only technology that can provide the world with all the energy it will ever need.

- Nuclear reactor waste-disposal problems are greatly relieved by the use of the proposed recycling and fast neutron reactors. Storage time is *greatly* reduced, along with the storage space required.

- GNEP essentially eliminates the potential problem of diverting reactor-grade materials to produce a nuclear weapon. Under GNEP, no nation will have to process its own nuclear fuel, except those few supplier-partner nations. The fuel and waste will be stored and managed by experienced supplier-partner nations. No more enrichment of uranium will take place anywhere in the world, except by experienced supplier-partnership nations that will supply fuel to all nations wanting to use nuclear energy for peaceful purposes.

▷ The goal is to have the fast neutron pilot plant with recycling facilities built and operating in less than 5 years. The United States and the world must expedite all regulatory approvals, and clear all political hurdles. A plan that calls for fast neutron commercial plants operating in 7 years should be put in place. Designs should be done during the pilot-plant stage. Such designs will buttress the 30-year transition plan and will cause the United States to spend less on expensive transition "bridges" (see Chapter Fifteen).

▷ There is one last detail. The world cannot realize the enormous benefits of GNEP, UREX+ recycling, or fast neutron reactors until it commits to them. In the United States a fast neutron reactor pilot plant must be built immediately and funded from the energy surcharges recommended. The United States should take the lead and help all nations become GNEP partners. Chapter Sixteen illustrates how the United States and other wealthy nations can easily afford to help poorer nations.

## *Chapter 11*

# CONCERNS ABOUT NUCLEAR POWER

All energy sources raise some concerns, and nuclear energy is no exception. Yet let's be clear: Nuclear energy *can* kill people, but fossil fuels *do* kill. That said, the nuclear industry is by far the safest *energy* industry in the United States and the world, and indeed it enjoys the best safety record of *all* industries.

The nuclear accidents at Three Mile Island and Chernobyl received widespread publicity. These accidents, discussed in detail below, largely shaped public attitudes toward nuclear energy. These accidents notwithstanding, the nuclear industry ably manages safety issues very well, but biased and often untruthful reporting often distorts public perceptions. Let's begin to clear up some confusion about these concerns.

- **Radiation.** This vague, frightening term needs explanation and context.

- **"China Syndrome."** This didn't happen at the Three Mile Island facility. If you worry that an out-of-control reactor could become a bomb, then you can relax. It cannot happen.

▶ **Proliferation.** The fear that a terrorist group or a rogue nation will get a bomb and detonate it, or threaten to, is a legitimate concern, particularly with all the bombs, highly enriched uranium, and weapons-grade plutonium lying around the world. The premise of the Global Nuclear Energy Partnership (GNEP) is that nations which pledge to forsake *nuclear weapons* but want *nuclear energy* should not be placed in the position of having to develop their own, indigenous reprocessing facilities, because such reprocessing could be diverted to producing weapons. Indeed, I believe that the entire issue of bombs can and will be effectively managed by the proposed GNEP.

▶ **Disposal of spent fuel and nuclear waste.** As one noted nuclear expert, declared, "The rational way to dispose of unwanted radioactive stuff is to put it in the silt at the bottom of deep ocean trenches, but that's too cheap to generate any lobbying pressure to modify the Law of the Sea Treaty. Thus, we fritter away billions on boondoggles such as Yucca Mountain." Even so, Yucca Mountain still provides a workable solution. Further, with recycling and fast neutron reactors, we only need about 20 percent or less of the disposal space we thought we would need. Storage time for waste from a fast neutron reactor is only 300–500 years compared to 10,000 years plus for waste from today's conventional reactors.

▶ **Uranium supply.** This becomes a non-issue with UREX+ recycling and fast neutron reactors.

▶ **Transportation of nuclear materials.** Such matters are well under control and have been for decades.

Let's examine each of these issues.

## RADIATION

Perhaps the most important misunderstanding about nuclear energy is radiation. For the public, "radiation" has an aura of mystery. It provokes primal fears of the unknown and vivid fears of mutation and painful death.

There have been many studies about nuclear radiation over the last 50 years. Much has been learned, and the world now effectively manages radiation exposure. For details I recommend Robert Morris' fine book *The Environmental Case for Nuclear Power* (Paragon Press) and Bernard L. Cohen's often-referenced *The Nuclear Energy Option* (Plenum Press). These are the most complete and coherent books on the safety of nuclear energy and radiation. Much of what immediately follows appears in Robert Morris' book.

### What We Know About Radiation

The nucleus of a radioactive atom—from, say, thorium or potassium—typical sources of background radiation—may release energetic particles (known as alpha or beta rays) or photons (gamma rays). These emissions are known as radiation. X-rays are also a form of radiation.

Let's consider radiation doses. Please see Figure 1.3 for comparisons.

*Extremely large doses* of radiation, as happened during the atomic bombings of Hiroshima and Nagasaki, can kill people. (However, far more people died from the effects of the blasts, which were qualitatively the same as that of conventional explosives, only bigger.) *Moderately high levels* of radiation may lead to the development of cancer and cause mutations in some of the lower animals, such as insects, but this has never been observed in humans. *At low levels,* such as most people are normally exposed to, there is good evidence that radiation apparently causes no net damage, but can actually stimulate cellular-repair mechanisms.

People cannot escape from radiation; we are constantly exposed to it. Each second about 15,000 particles of radiation strike each of us, totaling 500 billion collisions per person per year and almost 38 trillion collisions per person over an average lifetime. The source of almost all of this radiation is nature, not nuclear power plants. Much of this radiation comes from the soil and rocks, both of which contain a very small percentage of radioactive atoms, such as thorium, uranium, and their decay products. Food grown in the soil is also radioactive. An isotope of potassium (K-40) is the source of most of the radiation in our food and bodies. Finally, cosmic rays from space are a source of external radiation. Plainly, we cannot escape from radiation. We are all mildly radioactive. Nothing to worry about.

To distinguish the various levels of radiation, let's talk numbers. A *millirem* is one-thousandth (0.001) of a rem, a measure of the damage done to biological cells by radiation. A normal medical x-ray delivers about 95 millirem; a fatal dose of radiation is several million millirems. In the United States, the average person is exposed to about 250–360 millirems of radiation per year. This exposure seems to have little effect on our health. The International Commission on Radiological Protection sets 500 millirems as the maximum permissible annual dose an individual should receive. However, people who live in certain areas of India and Brazil, where the soil is rich in radioactive thorium, often receive over 1500 millirems per year without showing any increase in rates of leukemia or bone cancer.

About 130 of the 250 millirems we absorb yearly come from cosmic rays, the earth, and stone building materials. For example, if you live in a brick or stone building, you receive upwards of 30 millirems per year from the small quantity of radioactive substances in the minerals in the stone. And, although a person living at an altitude of 5000 feet in Denver, Colorado, receives 120 millirems per year from cosmic rays, one living at a lower altitude in Florida may receive only 35 millirems per year from the same source.

On average, humans absorb another 25 millirems from the food we eat. For the average person, this brings the total dose of radiation received from nature to about 155 millirems per year. An additional 95 millirems is added from human-made sources, largely medical x-rays.

Color television accounts for roughly 1 millirem. Thus, the average person is exposed to about 250 millirems of radiation each year. Little can be done to reduce this total aside from limiting the number of medical x-rays, which is obviously risky. Furthermore, there is no evidence that routine medical x-rays cause any adverse health effects.

## Low Levels of Radiation

Normal exposure is not worth worrying about. Although high levels of radiation can cause bone cancer and leukemia, there is no evidence that low levels of radiation cause either. Doses of radiation below 10,000 millirems are generally considered to be low-level doses.

The average American receives 250–360 millirems of radiation per year. Low-level exposure is less than 10,000 millirems. During the accident at Three Mile Island, people nearby were exposed to about 1.4 millirems.

Yet the International Commission on Radiological Protection wisely and conservatively recommends 500 millirems as an annual maximum.

In Denver, where radiation received from cosmic rays and minerals in the ground ranges from 100 to 150 millirems, depending on the elevation, the rate of both bone cancer and leukemia is *lower* than in New Orleans and San Francisco, where radiation from these sources averages only about 75 millirems. If lower levels of radiation caused very much bone cancer or leukemia, then the rates for these diseases should be highest in Denver, which has the highest background radiation due to its elevation. However, the opposite is true. Figure 11.1 shows the data.

## Figure 11.1. Radiation

| Location | Average Background Radiation from Natural Sources | Bone Cancer Rate Per 100,000 People | Leukemia Rate Per 100,000 People |
|---|---|---|---|
| New Orleans, Louisiana | 75 millirems | 2.80 | 6.90 |
| San Francisco, California | 77 millirems | 2.90 | 10.30 |
| Denver, Colorado | 100-150 millirems | 2.40 | 6.40 |

Source: *Radiation* by Martin D. Ecker, M.D. and other sources.

How does the radiation produced by nuclear power plants compare with other sources of radiation? U.S. federal law requires that radiation levels at the fence line of a nuclear power plant *not exceed 10 millirems per year*, which is only about one-tenth as much radiation as the average American annually gets from medical x-rays. Of course, hardly anyone lives this close to a nuclear power plant. The rest of us get less than 0.02 millirems of radiation per year from nuclear power plants. Unlike the standards set for air pollution, which are frequently violated, the standards set for nuclear power plants are closely monitored and almost never exceeded. Even during the highly publicized accident at Three Mile Island, people nearby were exposed to only about 1.4 millirems. Actually, power plants that burn fossil fuels actually emit more radiation than nuclear power plants, but this is not a concern since the radiation is so low. Figure 11.2 shows the levels of radiation from various sources.

## Figure 11.2. Various Radiation Doses

| Source of Radiation | Dose Received in Millirems Per Year |
|---|---|
| Average background in United States: cosmic rays, earth, and building materials | 130 |
| Average, all medical x-rays | 95 |
| Cosmic rays at sea level | 35 |
| Living in a brick house | 30 |
| Food, internal sources | 25 |
| Fallout from weapons testing | 3 |
| 3-hour flight in a jet | 2 |
| Watching color television | 1 |
| Maximum allowable level at the fence line of a nuclear power plant | 10 |
| All nuclear power plants: emissions over entire United States | Less than 0.02 |
| One coal-fired power plant: average within 20 miles | 0.10 |

Sources: various sources, most notably:

-Lawrence Berkeley National Laboratory < http://ap.lbl.gov/LBL-Programs/tritium/natural-dosage.html >.

-Reports from National Research Council, Advisory Committee on the Biological Effects of Ionizing Radiations, (National Academies Press, 1980).

-Princeton University's Open Source Radiation Safety Training at < http://web.princeton.edu/sites/ehs/osradtraining/backgroundradiation/background.htm >.

For comparison, note that occupants of a brick house receive 1500 times more radiation from the bricks and mortar than they do from nuclear power plants.

## Medical Studies

The atomic bombings of Hiroshima and Nagasaki exposed over 400,000 people to various levels of radiation. At least 300,000 people survived. A U.S.-Japanese Atomic Bomb Casualty Commission (renamed in 1975 the Radiation Effects Research Foundation) studied these survivors and their descendants over the next 30 years. Results indicate that exposure to massive doses of radiation causes three possible health effects: genetic mutations, radiation sickness, and cancer.

### Genetic mutations

Scientists have known since the early 1900s that large doses of radiation can produce genetic mutations in insects. However, no "excess genetic mutations" have been observed in any

> Power plants that burn coal emit more radiation than nuclear power plants.

children born to the Hiroshima and Nagasaki survivors. I wrote "excess" genetic mutations, because approximately 3 percent of all live births everywhere in the world show a mutation of some sort. That's just Mother Nature at work.

### Radiation sickness

Half of the people exposed to 400,000 millirems, which occurred during the atomic bombings, died. People exposed to over 100,000 millirems develop "radiation sickness," damage to the bone marrow which affects the production of white blood cells. Radiation sickness killed some people, but survivors saw their symptoms disappear within weeks or months. Below 100,000 millirems, radiation sickness usually does not occur.

*The level of radiation produced by all nuclear power plants in the U.S. is over 5 million times too small to cause radiation sickness.* The public need not worry about any radiation sickness caused by nuclear power plants.

## Cancer

Cancer, including leukemia, is the most serious health concern, because somewhat lower levels of radiation may cause cancer. (see Figure 11.3). Among people exposed to 20,000 millirems or less (which is a lot of radiation) at Hiroshima, where blast radiation produced especially dangerous neutrons, only the typical number of leukemia cases one normally finds in any population were present. Mother Nature at work again. At Nagasaki no excess leukemias were found at radiation levels below 100,000 millirems. To 1974, the carefully studied group of 80,000 Japanese survivors of Hiroshima and Nagasaki experienced only about 200 more cases of cancer than any randomly chosen population. Among this group, 8500 people received 100,000 to 600,000 millirems. However, 200 cancers among 80,000 people is 0.25 percent, a miniscule increase in view of the large amounts of radiation involved.

## Figure 11.3. Radiation Levels and Health Effects Among Survivors of Hiroshima and Nagasaki

| Radiation Dose in Millirems | Immediate Health Effects | Later Health Effects |
|---|---|---|
| over 1,000,000 millirems | Almost certain death | No genetic mutations at any level of radiation |
| 400,000 millirems | Nausea and fatigue; 50 percent chance of death | Excess cancers |
| 100,000 millirems | Radiation sickness; temporary changes in blood cell count | No excess cancers below this level at Nagasaki |
| 25,000 millirems | No medically detectable immediate effects below this level | None |
| 20,000 millirems | None | No excess cancers below this level at Hiroshima |
| 250 millirems (average exposure to radiation in the United States.) | None | No excess cancers likely at this level |
| 0.02 millirems (average radiation received from ALL nuclear power plants in the United States.) | None | None |

Sources: Various sources, including *The Nuclear Energy Option* by Bernard L. Cohen, Chapter Five.

A number of other important medical studies on the effects of radiation have been conducted on large groups of people, some of whom were exposed to radiation for years before its dangers were realized.

- 15,000 people who received large doses of x-rays in an attempt to treat arthritis of the spine

- 10,000 employees of the Hanford nuclear materials-production facility

- 36,000 U.S. hyperthyroid patients who underwent x-ray treatment

- 1600 Brazilians who live in an area of unusually high natural radiation

- Thousands of uranium miners

- 775 dial painters at a watch factory who, in the course of their work, ingested radioactive radium

These studies indicate that low-level radiation poses no threat to our health. In fact, some studies indicate that radiation below a certain threshold causes no damage and may be beneficial.

Although many Americans worry about radiation from nuclear plants, many studies confirm that the small amount of radiation emitted by nuclear power plants poses absolutely no danger to the public or the environment—NO DANGER. The widespread misconception is little wonder given the barrage of misinformation from various sources, including statements from respected, high-profile, poorly informed individuals. Actually, people have a better chance of being killed by the stress brought on by such irresponsible misinformation than from radiation.

According to Dr. Bernard Cohen, one source of radiation we should worry about has nothing to do with nuclear energy. Many fear that radon gas in our homes may cause lung cancer. While an average American gets 80 millirems per year from medical x-rays, radon gas in our homes can give the average American 200 millirems per year, many hundreds of times higher than typical radiation from a nuclear power plant. In some areas this radiation can be much higher. Public health authorities recommend carefully checking all houses for radon radiation.

## ACCIDENTS

According to the Bazley Institute, radiation has caused only 1 death in the U.S. commercial nuclear industry, and this death occurred at a reprocessing site, not at a reactor. This data goes all the way back to the 1950s, when our knowledge of nuclear energy was limited. Today, more than 50 years later, the nuclear industry has gained decades of operating experience and has gained much information from the few accidents that did occur. Since the United States has 103 active reactors, each having operated an average of roughly 30 years, the U.S. nuclear industry has a truly outstanding safety record, due largely to safe practices and regulatory compliance.

### Three Mile Island

The Three Mile Island reactor—located on an island in the Susquehanna River near Harrisburg, Pennsylvania—experienced a partial-core meltdown at 4:00 AM on March 28, 1979. The accident, which started in the *non-nuclear* portion of the plant, unfolded over 5 days. The accident was caused by technical malfunction and human error. The main cooling-water pumps failed, which prevented steam generators from removing heat from the reactor. Due to operator error, the back-up system did not operate. First, the turbine and then the reactor automatically shut down. Immediately, the pressure in the nuclear portion of the plant began to increase. In response, an automated pressure-relief valve opened, but it did not close again when the pressure decreased, although indicators

> People have a better chance of being killed by the stress brought on by such irresponsible misinformation than by radiation.

the operator was viewing showed that the valve had properly closed. Radioactive cooling water poured out of the stuck-open valve, which caused the reactor core to overheat. Finally, about 2.5 hours after the first malfunction, radiation alarms went off. Radiation reached 300 times normal levels. Still the operators did not know they were experiencing a loss-of-coolant accident—a partial core meltdown.

The accident was exacerbated by poor decisions from operators under stress. The operators were receiving misleading, incorrect, irrelevant, and excessive information. Notwithstanding the magnitude of the accident, the partial meltdown did not create a "China Syndrome" event, nor did it cause massive deaths. A severe core meltdown occurred, the most dangerous kind of nuclear power plant accident. Even after a meltdown of one-third of the fuel, the reactor vessel functioned as designed, maintained its integrity, and contained the damaged fuel. Some feared a breakdown of the containment walls, but this did not happen. People living within 10 miles of the plant received about the same radiation a person would receive from a chest x-ray, and the highest amount of radiation any single individual received was equal to about one-third of the average background level of radiation received by a U.S. resident in a year.

Although the accident was a frightening experience, the design of the containment system worked—a remarkable and comforting conclusion. Experts were relieved by the outcome, while many less-informed activists remained unconvinced.

### After the Accident

The accident at Three Mile Island caused near panic. The media made sure that every catastrophic possibility, and even some impossibilities, were repeatedly reported. The media inaccurately extended these doomsday predictions to the *entire* nuclear industry. Yet the entire industry did not have such troubles. No one can blame members of the media for doing their jobs as they saw fit, but more balanced reporting would have provided a much greater public service. As is so often the case, media outlets typically quoted only activist groups with full-time press release professionals selling their points of view. Hello, Union of Concerned Scientists.

Almost 70,000 anti-nuclear activists marched on Washington to demand that all nuclear power plants be shut down. While these groups can assume all kinds of righteous authority, they take no responsibility for their actions and are short on viable alternatives or solutions.

The reactor vessel functioned as designed, maintained its integrity, and contained the damaged fuel.

The accident put the nuclear industry on the ropes. Activists went in for the knockout punch. Their success was a victory for fossil fuels. As a result, many of us suffer with respiratory diseases and rising medical costs. It was also a victory for other national catastrophes such as acid rain, acidic lakes and oceans, mercury contamination, and other environmental woes. The public now knows the cost of burning coal and fossil fuels is incalculably expensive. While new construction of nuclear plants had stopped even before the Three Mile Island accident, the incident and the outraged public reaction threatened to kill the nuclear power industry in the United States. Some utility companies tore down billions of dollars worth of nearly finished nuclear power plants.

The world, and the United States in particular, pays a stupendously high price for the successful efforts by the media and special interest groups to demean and demonize the nuclear industry. We know that nuclear energy has proven itself safe, secure, reliable, affordable, and efficient in the United States, Europe, Japan, and many other parts of the world, not withstanding the very serious proliferation issue (discussed elsewhere in this chapter).

## Chernobyl: The Good and the Bad

Seven years later, the worst nuclear accident in history took place at Chernobyl, Russia, on April 25, 1986. The accident dispersed about 5 percent of the radioactive core into the environment as airborne dust. This airborne dust is especially dangerous because it can affect a very large area. A cloud of radioactive fallout drifted over parts of the western Soviet Union, Eastern Europe, Scandinavia, the United Kingdom, and the eastern United States. This accident was the result of a terribly flawed reactor design operated at the time by poorly trained personnel.

The Chernobyl accident occurred during an electrical engineering experiment conducted with no reactor experts in attendance. A chain of operating misjudgments took place, with little regard for several strict operating rules. Finally, about 2 hours after the experiment started, a computer warned that the reactor was unstable, unsafe, and should be shut down immediately. Yet the operators ignored the warning. Nobody knows why, and we will never know, because they died in the accident. The real tragedy, according to the designers of

> The world, and the United States in particular, pays a stupendously high price for the successful efforts by the media and special interest groups to demean and demonize the nuclear industry.

the plant, was that the operators knew the reactor was dangerously unstable under the test conditions proposed, but this information was not shared with the engineers conducting the experiment.

The operators tried to control the problem by inserting control rods, but the effort was too late. Some fuel tubes had already broken open, and the chain reaction went out of control. If this happened in a U.S.-designed light-water reactor, the loss of the water moderator would have stopped the chain reaction immediately, although there could be damage to the core, as occurred during the accident at Three Mile Island. Instead, Chernobyl's graphite moderator, by design, is permanently in place, and accommodates the chain reaction. During the Chernobyl accident, the graphite moderator caught fire. In a short time reactor power was many times greater than the amount the structure could handle. A steam *non-nuclear* explosion blew the top off the reactor. It is believed there was a secondary hydrogen gas explosion. The speculation was that the hydrogen gas was generated in a chemical reaction when water or steam came in contact with the hot zirconium fuel cladding. Radioactive material continued to spew into the environment for days.

Heroic firemen put out the graphite fires in the 4 hours after the reactor disintegrated. Many firemen later died from gamma radiation.

Helicopter pilots dropped fire retardant materials into the reactor in heroic efforts to stop the chain reaction. As a result, the radioactive discharge dropped to a low level about 12 days after the accident. Many of the pilots died. During this entire tragedy there was no shortage of heroes.

At the reactor site 31 people died. Some were exposed to more than 1 million millirems.

This accident caused the press and others, particularly in the United States, to strongly question the nuclear industry. The media led people to believe that equipment was not and could not be properly designed, that people in the industry were simply not as competent as they should be, and that nuclear energy costs were simply too high. Also, the coal industry wasted no time in pointing out that since the United States has plenty of coal, who needs nuclear energy? Further, some people thought this kind of accident could happen in the United States. It can not happen.

Let me be as clear as possible about the last point: *A Chernobyl-like accident cannot—cannot—happen in the United States* because the reactors are designed differently, and a containment building is included in U.S. designs to prevent radioactivity from escaping into the environment in case of an accident—exactly the problem that happened at Chernobyl. Soviet/Russian officials chose a less-safe design for the Chernobyl reactor because the facility was initially designed to produce Plutonium-239, primarily for making bombs and secondarily for producing electricity.

Although the Chernobyl accident could not have been worse and the facility lacked a containment structure to prevent radiation from escaping, the radioactive intensity of the fallout was so low that no adverse health effects have been documented, except for reports of excess thyroid cancers near the accident site, mostly in children.

Before the accident, some prominent anti-nuclear activists issued many, scary predictions about a nuclear future. Perhaps the most noted of these activists, Ralph Nader, said, "I don't think that a society can endure the disaster of one major meltdown.... I really don't think that our country can tolerate the trauma of a couple of hundred thousand people dying all at once in one place, and many more dying over a period of time from cancer, leukemia, mutations and what have you." Many others predicted similar consequences. Such fears are overstated and misplaced. While tragic, the Chernobyl accident had a small virtue. It actually removed much of the fear of an Armageddon-type disaster resulting from a reactor meltdown.

> A Chernobyl-like accident cannot happen in the United States because U.S. reactors are designed differently and include a containment building.

### Since the Accident

It is difficult to tally accurately the number of deaths caused by the Chernobyl accident, because most of the expected, long-term fatalities, especially those from cancer, have not yet actually occurred (or may never occur) and will be difficult to attribute specifically to the accident. However, a 2005 report from the International Atomic Energy Agency attributes 56 direct deaths to the accident (47 accident workers and 9 children with thyroid cancer), including the 31 that died at the reactor site. The report estimates that as many as 4000 people may ultimately die from long-term, accident-related illnesses. This figure is much lower than the massive numbers of deaths initially predicted. Also, the figure of 4000 is more than 1000 times less than the number of deaths caused by the burning of fossil fuels, particularly coal, every year. Loss of life from this accident is about the same as the risk of contracting cancer from drinking 3 cups of coffee a day, 7 times less than from car accidents, and 1000 times less than from smoking.

So far, there has been no increase in leukemia, congenital abnormalities, adverse pregnancy outcomes, or other radiation-induced diseases in the contaminated areas. The loss of life as a result of this accident, while tragic, was still far less than the infamous fossil fuel-induced Smog of 1952, which enveloped London and killed 12,000 Londoners.

Since the Chernobyl reactor did not have a containment structure like those in U.S. nuclear plants, the Chernobyl facility was an accident waiting to happen. A containment structure would have prevented radioactivity from escaping outside the plant, and there would have been no deaths. The accident was exacerbated by bad reactor design, the incompetence of local administration, and lack of proper equipment and planning. The reactor crew could not even determine the radiation levels in various parts of the reactor building—the meters were wrong or were broken.

The International Atomic Energy Agency blamed the Chernobyl accident on the reactor's design.

In 1992, almost 7 years after the accident, the International Atomic Energy Agency, the U.N. nuclear watchdog, attributed the root cause of the accident to the reactor's design and not to operator error. In 1986, at the time of the accident, analysts cited operator error as the likely principal cause of the accident.

In September 2005, news agencies around the world were reporting recent conclusions from the International Atomic Energy Agency: "Fewer than 60 deaths have been directly attributed to radiation released by the 1986 Chernobyl nuclear power plant accident, and the final total could be thousands less than originally believed." Susanna Loof, writing for the Associated Press, reports: "The death toll could reach 4000, but [study] chairman Dr. Burton Bennett said that previous death tolls were inflated, perhaps to attract attention to the accident, to 'attract sympathy.'" In any event, that 4000 number is based on a theoretical model of radiation damage (linear extrapolation from high-dose data) that has no empirical support.

This accident, however unlikely, could happen again in Russia since 12 identical or very similar reactors still operate there. These reactors have been made much safer, but still with no containment building. The particular design of this so-called RBMK reactor has serious design flaws regarding the management of the cooling water and control rods. However, the control rod problem has reportedly been resolved.

### History and the Safety of Nuclear Reactors

Senator Pete V. Domenici, in his fine book *A Brighter Tomorrow*, cites a 1998 study by Stefan Hirschberg and associates of the Paul Scherrer Institute in Switzerland. The study analyzes accidents in energy-related industries from 1969 to 1996, the time period that includes the accidents at Three Mile Island and Chernobyl. The Institute reports 31 deaths from the Chernobyl accident and none from Three Mile Island. The Institute's database of 13,914 severe industrial accidents (across all industrial sectors) includes 4290 energy-industry-related incidents. The comprehensive data analysis found that *of all energy sources—including coal, oil, natural gas, liquefied petroleum gas, nuclear, and hydropower—*nuclear power is significantly safer than all other forms of generating energy. Natural gas, the next safest compared to nuclear, has a fatality rate 10 times higher than nuclear—10 times higher.

According to Senator Domenici, the study from Hirschberg and colleagues concludes that for each terawatt-year of energy use, the number of fatalities worldwide was 8 for nuclear, 85 for natural gas, 342 for coal, 418 for oil, 884 for hydro, and 3289 for liquefied natural gas. A *terawatt-year* is an extremely large quantity of energy, much more than the yearly electrical energy consumed in the United States.

## Earthquake Damaged Japanese Reactor in 2007

A very strong earthquake in Japan registering 6.9 on the Richter scale severely damaged peripheral buildings and equipment at a large nuclear electricity generating plant, although the reactors themselves were undamaged. Some mildly radioactive water leaked from the plant, but harmed nobody. Early media reports predicted dire consequences.

Simultaneously, official sources reported that, in spite of the severe damage, the safety systems operated as expected and the facility was well under control shortly after the earthquake subsided.

## Lessons Learned from Reactor Accidents

The Chernobyl accident did not really teach us much, except that the reactor was poorly designed and that a containment structure is absolutely necessary. However, the Three Mile Island accident taught valuable lessons. Existing and future plants were made safer and more secure, and permanent and sweeping changes improved how the U.S. government's Nuclear Regulatory Commission regulates the industry. *Any plant built in the future will be safer than any built in the past, even though the industry has enjoyed an exemplary safety record that is the envy of the entire energy and chemical industries—and indeed of the industrial world.* The experience of the nuclear reactor damaged in the Japanese earthquake vividly illustrates the integrity, safety, and durability of new nuclear plants.

> Natural gas, the next safest energy source compared to nuclear, has a fatality rate 10 times higher than nuclear.

## PROLIFERATION and GLOBAL STOCKPILES of NUCLEAR WEAPONS

*The 5 major nuclear powers*—the United States, Russia, Britain, France, and China—*have more than 30,000 nuclear weapons in their war arsenals.* More than 125,000 nuclear bombs have been built since 1945. Most existing bombs are in the United States or Russia. The United States has over 10,000, down from 70,000. Russia has somewhere between 9000 and 18,000, down from over 50,000. Britain has approximately 180, down from 350; France maintains approximately 350, down from over 1200; China has approximately 400. U.S. defense agencies predict that China may increase the number of warheads aimed at U.S. targets from 20 to 75–100.

Other countries have nuclear weapons, too. Israel, while it has neither confirmed nor denied possession of nuclear weapons, is believed to have approximately 200 warheads. India and Pakistan have approximately 100 between them, and experts believe these countries will increase their stockpiles. As of this writing, North Korea is the newest member of this getting-to-be-not-so-exclusive club. Iran seems on the way to becoming the next member, and I believe many more will follow without concerted international action to bring the situation under control.

Countries claim to need nuclear weapons for self-defense and for a retaliatory capability—that is, the ability to inflict nuclear destruction on those who might attack first. This is a balance of terror or a balance of insanity that experts call "mutually assured destruction." The rest of us call it M.A.D. With more and more countries joining the club, where will it stop? New weapons, like new toys and new tools, tend to get used.

## The Obvious Solution

If world leaders really want to make the world a safer place, then they must begin to eliminate the nuclear arsenals of the world. There simply is no other solution. The United States and Russia should lead the effort. Short of that, the arsenals must be reduced to a minimum, maintained under the direction of the United Nations or the proposed GNEP, and used only to disarm any cheating nation. And short of that, nations that keep nuclear weapons must agree to possess only a minimal deterrent arsenal. For example, I don't think the United States needs more than 300 warheads. No nation, however honorable, should have the ability to annihilate the world.

Non-proliferation treaties help as much as a bandage "helps" a deep gouge—a temporary measure for symptoms, but not for underlying causes. Countries can and have decided to withdraw from the Nuclear Nonproliferation Treaty, as North Korea did when it chose to make nuclear weapons.

Objecting to nuclear energy based on the proliferation argument makes no sense, because the world is already awash in nuclear weapons and nuclear power-plant waste. Further, governments can build nuclear weapons without building nuclear power plants. The United States built bombs in the 1940s and did not have a nuclear power plant until the 1950s. Eighteen countries already own uranium enrichment capabilities, the technologies and material to create "bomb stuff."

The only real answer is a 3-pronged approach: the total or near total elimination of nuclear weapons, the deployment of fast neutron reactors, and the proposed GNEP. This approach would hasten a safer world and abundant energy for all nations. It would also save countless billions of dollars currently spent to maintain and update these weapons. Governments should spend that money on education, health, and averting the energy crisis.

## Can Rogue Nations and Terrorists Make Weapons From Nuclear Material Used to Generate Energy?

The short answer is an emphatic no. The long answer is no, no, no. It would be infinitely easier to acquire a credible nuclear weapon by buying one on the black market or by buying or stealing some Plutonium-239 than to try to make a bomb from reactor-grade waste, particularly the waste from a fast neutron reactor. The crucially essential point: *Reactor-grade* material is not the same as *weapons-grade* material.

Governments can build nuclear weapons without building nuclear power plants.

To be sure, many glaring, troubling proliferation problems exist, but managing the spent fuel from power reactors is not one of them. Creating weapons-grade Plutonium-239 is a fairly detectable and expensive endeavor. For example, observers detected such efforts in Iraq, North Korea, and elsewhere. Once a government invests in the necessary facilities, a power-generating reactor must be run in brief cycles designed specifically to extract weapons-grade Plutonium-239. The extracted plutonium must then be enriched in an additional process

(PUREX). Such extraction and processing is enormously expensive. To minimize proliferation risks and costs for partner nations, advocates of GNEP want to centralize reprocessing in a few, secure locations in the world—the supplier-partner nations—and create a system of shared use, distribution, and recycling among recipient-partner countries.

Whether new nuclear power plants are built or not, any nation with reasonable technical capabilities and adequate resources can make a nuclear bomb, and there is nothing the United States or any other nation can do about it except exert political pressure or military action. Yet the ability of the United States to exercise political leadership or exert political influence has waned. U.S. actions and rhetoric squandered our once formidable political position in the world. A vain waste. In my old ghetto neighborhood only fools unnecessarily provoked others, especially enemies. Why provoke a potential future ally or spark vengeance?

### Rogue Nations

Since nations have considerable financial and technical resources, they could acquire the means to build a nuclear weapon. The effort is expensive and requires non-trivial interdisciplinary technical skills. Let's see how *nations*, especially rogue nations, could acquire nuclear weapons, listed in order of possible success.

1. **Steal or buy a nuclear weapon.** This is not easy because of security, size, and weight. Such a weapon would likely be detected in transit. The smallest weighs approximately one-half ton, but most are much larger. A stolen bomb probably could not be detonated because of built-in safeguards. However, if a nation had the requisite technical capability, the stolen weapon could possibly be disassembled and reassembled to work.

2. **Buy or steal highly enriched uranium or weapons-grade Plutonium-239.** This is a troubling prospect

because so much of this stuff sits around the world. Numerous reports suggest some of this material has already been sold clandestinely, much of it from Russia shortly after the collapse of the Soviet Union. Of course, after acquiring such material, one still faces the non-trivial task of fabricating a usable weapon.

3. **Buy uranium ore and equipment to enrich the content of Uranium-235 to a concentration sufficient for bomb-making, and employ many scientists in various disciplines to fabricate a bomb.** They don't need a nuclear reactor, but they do need considerable money and technical talent to be successful.

4. **Build a light-water reactor and use it to make weapons-grade Plutonium-239.** Run the reactor for short periods of time (30–60 days), then extract and concentrate the Plutonium-239 until one has enough to fabricate a bomb. The longer the reactor runs, the more the Plutonium-239 is subjected to neutron bombardment. The more bombardment, the heavier the plutonium becomes, converting to Plutonium-240, then Plutonium-241, then Plutonium-242. These isotopes are lousy for making a bomb. Also, a nation would still need significant financial and technical resources to fabricate a nuclear weapon.

5. **Steal, buy, or use "waste" (used, spent fuel) from a light-water reactor—the material destined for storage in Yucca Mountain or recycled for use in fast neutron reactors.** This nasty material is easily detected and difficult to handle and transport. The spent fuel is very radioactive and is in 6- to 12-foot rods weighing hundreds of pounds each. Only an idiot would try to use it to make a weapon, even though theoretically one might be able to make a low-grade device if the builder doesn't fatally irradiate or otherwise kill himself in the process. For the

danger and effort, the almost certain result would be the construction of a weapon that won't work.

6. **Steal, buy, or use some fuel or waste from a fast neutron reactor.** If creating a nuclear weapon from spent fuel from a light-water reactor is essentially impossible, then doing so with the recycled fuel for a fast neutron reactor is doubly impossible because the fuel is much more radioactive. Also, the waste material from a fast neutron reactor contains only the faintest, trace amounts of plutonium, so serves no purpose as a weapons component.

Only the first four possibilities listed above are conceivably viable paths to making nuclear weapons.

## Terrorists

While just about any nation could build a nuclear weapon, it would be extremely difficult for any terrorist group to acquire or steal weapon-making materials and then fabricate a workable weapon. Although there is a *remote possibility* that a terrorist group could acquire a bomb, bomb-grade uranium, or plutonium, and then devise a weapon, it is virtually impossible for them to access reactor-grade material and fabricate a credible weapon.

> It would be extremely difficult for any terrorist group to acquire or steal weapon-making materials and then fabricate a workable weapon.

The 6 scenarios above apply. However, scenarios 3 and 4 seem out of reach for terrorist groups, either economically, technically, or both. Also, terrorist groups confront the added burden of detection and interdiction. Scenarios 1 and 2 above seem the most reasonable for a terrorist group, but the daunting task of actually making a usable weapon and the strong possibility of detection are formidable deterrents. Scenarios 5 and 6 above—making a nuclear weapon from spent fuel and nuclear-plant waste—offer no hope for a terrorist group, and only fools would try. Besides, there are easier ways to destroy life.

There is only a small, scant possibility that a sub-national or terrorist group—even one as well organized and financially well-to-do as Al-Qaida—could develop its own nuclear weapons without a lot of outside help in obtaining the necessary nuclear material and the necessary technologies to fabricate the device and still avoid detection. However, to repeat, there exists a remote risk of well-organized terrorist groups somehow acquiring a fully assembled nuclear weapon from a friendly nuclear state or somehow acquiring sufficient weapons-grade material to significantly reduce the barrier to making one. I believe the proposed Global Nuclear Energy Partnership is the answer to these potential problems.

If GNEP is implemented soon, then everybody today and in the future could feel safe from the proliferation of nuclear weapons. GNEP proposes to manage fuel and waste worldwide. The nearly total elimination of existing nuclear weapons, and weapons-ready uranium and plutonium, would provide even greater worldwide security.

Building more nuclear power plants will not make the proliferation issue any worse since the opportunities to make a bomb cited above already exist all over the world.

### Dirty Bombs

A dirty bomb, sometimes called a *radiological dispersal device* (RDD), refers to a bomb that combines some radioactive material with a natural explosive to create a device designed to disperse radioactive material over a large area. Since the attacks of 9/11 on New York City and Washington, D.C., many people around the world have been worried about such a bomb. However, a dirty bomb is unlikely to cause many deaths, except for those deaths from the initial conventional explosion. In fact, in the 1960s the U.K. Ministry of Defense decided that using conventional explosives in place of radioactive material would be more destructive. Nevertheless, the prospect of a dirty bomb and radioactive exposure causes great fear, perhaps its greatest threat.

Radioactive material for a RDD could come from millions of sources used for medical and industrial purposes around the world. Reactor-grade radioactive material would be the most difficult to acquire, because of its size, inherent properties, and accompanying security systems.

To build a RDD, handlers must overcome many logistical hurdles. The material must be sufficiently radioactive, the material must be safely transportable, and the material must be dispersible. Satisfying all three requirements is difficult. In any case, assembling and transporting a RDD bomb without severe radiation damage and possible death to the perpetrators would be extremely difficult, too. For example, if the radioactive material were properly shielded to prevent contamination during assembly and transport, then the bomb would be very difficult to transport, easier to detect, and much less effective if detonated. If the material were not shielded, then transporting the material and assembling a weapon would be very difficult before the handlers succumbed to radiation sickness and perhaps death.

No radioactive dirty bomb has ever been detonated, although some groups have assembled lethal *chemical* weapons. Since the world has no experience to draw upon, many fear the unknown. However, the experiences at Chernobyl, the experiences at some accidents involving radioactive material, and the results of several analyses indicate that RDDs will neither sicken nor kill many people.

## NUCLEAR WASTE DISPOSAL

Operation in the United States of all 103 nuclear reactors, with no recycling, generates about 2000 tons of nuclear waste per year. This waste would occupy only the space of a medium-size house. However, the storage space must be much greater because of the heat generated by the waste—thus Yucca Mountain.

The proposed storage site at Yucca Mountain is insufficient to support vast deployment of light-water reactors. To comply with current legislation, Yucca Mountain will be filled with the waste from current light-water reactors within a decade. By coupling recycling to fast neutron reactors, the storage capacity of Yucca Mountain would be increased five fold. Moreover, the storage period is reduced from 10,000 to 300–500 years.

## ABOUT URANIUM

### Natural Uranium

The use of some forms (oxides) of uranium dates back to the second century AD, when it was used to add a yellow color to certain glasses made in Italy. Credit for the discovery of uranium goes to Martin Klaproth, a German scientist, in 1789. It was named after the planet Uranus, which was discovered at about the same time. The pure metal, which is silvery-white in color, was isolated in 1841, and by 1850 it was being used commercially in the glass industry. It wasn't until about 1900 that a French physicist, Henri Becquerel, discovered that uranium was radioactive.

> GNEP is one of the genius ideas of the twenty-first century.

When refined, uranium is weakly radioactive, softer than steel, and almost twice as dense as lead. Natural uranium consists almost entirely (99.28 percent) of the isotope Uranium-238, and almost all the rest is Uranium-235 (0.7 percent). Only Uranium-235 is used to fuel light-water reactors. To make fuel for light-water reactors, the uranium ore must be "enriched" to raise its Uranium-235 content to about 4 percent.

### Depleted Uranium

The uranium in the leftover "tailings" from the enrichment process is called "depleted uranium." While no longer useful for thermal-reactor fuel, the depleted uranium tailings still contain more than 80 percent of the energy that was in the original ore. Since that energy can be accessed with fast neutron reactors, the depleted uranium constitutes a very large energy resource. In fact, the energy in the depleted uranium waste already on hand in the United States far exceeds the energy in the coal reserves still in the ground.

Since it is so very dense, depleted uranium is sometimes used to clad military vehicles. After many studies and considerable research, there was no evidence to support any concern whatsoever for the safety or health of military personnel or anybody else.

## TRANSPORT of NUCLEAR MATERIALS

Many believe the transportation of nuclear materials is the most vulnerable link in the chain of nuclear-material handling. This has posed no problems in the past, and, using fast neutron reactors, will pose even less of a problem in the future. Also, if deemed necessary, the nuclear industry could easily afford armed protection of materials in transit and at other vulnerable sites.

## BOTTOM LINE

- Any new nuclear plant will be safer and more efficient than any built in the past. The United States has over 30 years of exceptionally safe operating experience—the safest of any industry.

- Storage and proliferation problems are greatly reduced by using fast neutron reactors and the new UREX+ recycling process, which was designed to be proliferation-resistant. Waste-storage space will also be greatly reduced and long-term waste-storage time would be dramatically less—300–500 years instead of 10,000 plus years.

- The United States must expedite the building of the fast neutron reactor pilot plant. Start immediately, and be operational in 4 years.

▷ Fast neutron reactors must be fast tracked and the GNEP should be aggressively pursued in cooperation with other nations. GNEP in my opinion will one day be deemed the most enlightened idea of the twenty-first century.

It is worth repeating that nuclear energy (fast neutron reactors) is *inevitable* if the nations of the world are to maintain their economies and provide energy for future generations.

## Chapter 12

---

# WATER

### ANOTHER CRISIS AVERTED?

A discussion of water is essential to any discussion about energy. Huge amounts of water are needed to produce electricity and some fuels. Of course, huge amounts of water are also necessary to sustain all life on earth and to modulate the earth's climate. Yet the world is running out of uncontaminated fresh water. Already a full 20 percent of the world's population does not have access to uncontaminated drinking water. So here's the crisis: To remedy the shortage of water by creating fresh water from sea water—a process called desalinization, which removes salt from ocean water—requires a lot of energy, yet burning fossil fuels for energy strongly contributes to the contamination of water. This is a catch-22. The horns of a dilemma. World shortages of clean, fresh water is a first-class, three-alarm, condition-red, world-scale crisis—a crisis that can be averted only by changing human behavior (good luck with that) and/or by using renewable energy sources to desalinate seawater.

Without food a human can live for 30–40 days, but without water one can live only 4–5 days. Does the world need desalinization plants to produce clean, fresh water? Absolutely.

That means the water crisis is also an energy issue. Access to abundant, inexpensive, non-polluting energy is the *only* solution to the water crisis.

## GRASPING the PROBLEM

There have always been conflicts and even major wars over access to water. In fact the word "river" is derived from the Latin word *rivalis*, which implies rivals in conflict over the same stream. When not enough water is available, people can pray for rain, start digging wells, or go to war.

There is no agreed standard for the amount of water a person needs every day, but experts put the minimum at about 13 gallons. For personal use Americans consume about 140 gallons per person per day—far more than anybody else on the planet, about 9 times more than the average African. The Japanese and most Europeans consume less than half that amount. China consumes 23 gallons per person per day, yet people in Mozambique use less than 3 gallons per person per day.

> Already a full 20 percent of the world's population does not have access to uncontaminated drinking water.

Water, sometimes thought of as the ultimate renewable resource, covers over 70 percent of the earth's surface. Over 97 percent of the earth's water is in our saltwater oceans. Glaciers and polar ice caps hold over 2 percent of the total water, and fresh water in lakes, streams, aquifers, soils, and the atmosphere hold less than 1 percent of the total. For every gallon of fresh water in the world's lakes and streams, 50 gallons sit in the ground in large aquifers. Water is always in motion through evaporation and precipitation and through the take up and dispersal of water by plants and animals. However, the ability of the water cycle to provide adequate fresh water faces three monumental problems: rising demand, falling supplies, and deteriorating water quality.

The *demand* for fresh water from the rapidly growing world population and growing industrialization far exceeds the *supply* and the ability of nature to provide it. Throughout history many of the great cities on earth have grown and been sustained by large rivers, yet today no river can satisfy the demands of the world's largest cities. Some places already face a water crisis, even at present population levels. We recognize the tragic images, if not the names and locations: Darfur, Ethiopia, Somalia. The scope and intensity of regional and local water crises will accelerate as the world adds the equivalent of 2.5 more Chinas to the global population in the next 30–50 years. In 30 years water use is predicted to increase by approximately 50–60 percent—far faster than population growth.

What of water *quality?* The burning of fossil fuels releases large volumes of gases that form acids and emit mercury and other pollutants into the environment. These pollutants end up primarily in the earth's water. Other human activities have damaged water as well: The extensive use of fertilizers, herbicides, pesticides, animal and human waste, and other chemicals, as well as excessive irrigation, have all played a role. Add mismanagement and wasteful practices to this mess, and you have an almost irreversible problem—part of the perfect storm.

## The Struggle for Safe Water

*Clean water has had as much or more to do with human health than the development of miracle drugs. The lack of clean, fresh water is as devastating as famine, plague, epidemic, and war.*

- In the last decade more children died from diarrhea than have been killed in all the wars since World War II.

- 40,000 children under age 12 die every day from diarrhea, an illness worsened by lack of fresh water.

- ◖ 6000 children under age 5 die every day specifically because of unsafe water and poor sanitation.

- ◖ 1.1 billion people in developing countries have inadequate access to safe drinking water and have never seen a toilet.

- ◖ Water-related diseases are the single largest cause (50 percent) of human sickness and health-related death in the world.

- ◖ Many women and children spend up to 6 hours per day fetching water.

- ◖ Many families in the developing world often spend up to 25 percent of their income to purchase water and carry it home.

Water-related diseases are the single largest cause (50 percent) of human sickness and health-related death in the world.

The United Nations designated 2000–2009 as the "Water for Life" decade and established in 2000 eight decade-long Millennium Development goals aimed at eliminating the world's most desperate poverty. One goal seeks to cut in half the number of people without access to clean drinking water. Another goal sets a similar standard for improving access to sanitation facilities. As of 2007 there is no chance of meeting either goal, and they never will without abundant, relatively inexpensive energy.

### Irrigation Miracle? Yes, But...

Which is more important? Drinking water or eating? Amid the growing frenzy over access to fresh water, the world's farmers and agricultural businesses must continue to irrigate crop lands because of the strong link between irrigating food crops and alleviating poverty. In many countries, *a full 60–70 percent of all water use is for irrigation*, which helps feed the world's surging population.

Quite simply, the world trades fresh water for food.

According to Sandra Postel in *Pillar of Sand*, "some 40 percent of the world's food comes from irrigated cropland." A stable supply of food for the world's hungry mouths now depends on an increasing global water deficit. According to the Worldwatch Institute, over-pumping for irrigation is sucking out too much of the world's underground fresh water, leaving insufficient amounts for health, sanitation, industry, and future generations. As a result of irrigation, for large portions of the year many major rivers now run dry—including the Yellow River in China, the Indus River in Pakistan, the Ganges in South Asia, and the Colorado in the American Southwest. Irrigation's heavy demands for water also damage the health of the aquatic environment by shrinking wetlands, reducing fish populations, and pushing species toward extinction.

> World shortages of clean, fresh water is a first-class, three-alarm, condition-red, world-scale crisis.

At the same time, the productivity of irrigated lands is in jeopardy from the buildup of salts in the soil from over-pumping water and from the growing diversion of irrigation water to cities. As population grows, the amount of irrigated land per person shrinks.

To meet food needs, many countries—especially those low on water and viable soil—import grain. Indeed, a landmark study by Henery Kindall and David Pimentel reports that only 2 countries in the world are major food exporters: the United States and Canada, and these countries irrigate extensively. Janet Raloff reported that in 2002, "for the third year in four, world per capita grain production fell … [and] last year's per capita grain yield was the lowest in more than 30 years." Raloff, quoting from *Vital Signs 2003*, a joint publication of the Worldwatch Institute and the United Nations Environment Programme, adds that "production of the world's three major cereals fell in absolute terms in 2002." In May 2006, Stewart Wells of Canada's National Farmers Union (NFU) predicted "a calamitous shortfall in the world's grain supplies in the near future… [because] in five of the last six years global population ate significantly more grains than farmers produced." The NFU adds, "North America's industrial-style agricultural system [with its heavy reliance on irrigation and chemical pesticides, herbicides, and fertilizers] is a really bad idea and may be the worst on the planet." In the "Hope" section of the report, the NFU encourages progress on all relevant fronts to assure that "poor regions [will be] *using renewable energy* to power a new, and clean, era of prosperity" (emphasis added).

## A SURVEY of PROBLEMS

The United Nations World Water Assessment Program reports that the Asian continent, which supports 60 percent of the world's population, has only 36 percent of the world's freshwater resources. North and Central America, with 8 percent of global population, enjoy 15 percent of the world's freshwater. Others? Europe has 13 percent of global population and 8 percent of freshwater. Africa has 13 percent of the population and 11 percent of water. Then there is South America with about 6 percent of the world's population and 26 percent of the freshwater—an enviable combination.

> Quite simply, the world trades fresh water for food.

### United States

The United States has a bigger problem than you might think. According to the U.S. Geological Survey, the United States uses about 350 billion gallons of fresh water every day or almost 128 trillion gallons per year. Americans use a total of 15 trillion gallons per year for personal use. Where does the water come from?

Over 90 percent of the fresh water in the United States is underground. Approximately 42 percent of the water used for irrigation comes from underground water. The largest aquifer is the Ogallala, which provides irrigation water for about 20 percent of the irrigated land in the nation. The aquifer is being depleted far faster than it is being replenished by rain. The Ogallala lies under 8 states, from North Dakota to Texas, and as it is being depleted, it is also getting polluted. The amount of water extracted from the Ogallala aquifer each year is approximately equivalent to the flow of 20 Colorado rivers.

Irrigation water should be better managed or allocated by state and federal authorities to allow slow replenishment of aquifers for use by future generations. For example, state and federal laws should protect the Ogallala and other depleting aquifers. Such protection will require major political intervention since many will want to dip their political straws into this great aquifer. Portions of the Ogallala aquifer have dropped 100 feet since the 1940s, according to a study by the National Academy of Sciences mentioned in an editorial in the *Minneapolis Star Tribune* on November 14, 2007. One way to protect the aquifers is to produce irrigation water by using a renewable energy source.

According to the U.S. Geological Survey, about 83 percent of water used in the United States goes to producing electricity (50 percent) and irrigating crops (33 percent). Electricity production uses over half of all fresh, surface water in the United States, while irrigation consumes the largest total amount of fresh water, including water from lakes, reservoirs, ground water, and aquifers. In 2000, over 40 percent of irrigation water was drawn from underground sources. Much water is used for cooling energy-producing equipment then returned to its source, making it available for other uses. Personal use from publicly supplied water consumes about 11 percent; aquaculture and mining use about 3 percent of the total water used in the United States. Industrial use consumes the rest.

Water conservation is always appropriate because there's a cost to every drop of water that we use. Yet even if we all stopped bathing, drinking, and cooking with water, the reduced use would only modestly affect overall use of water in the United States.

> The amount of water extracted from the Ogallala aquifer each year is approximately equivalent to the flow of 20 Colorado rivers.

Why not build more dams to create more reservoirs? They make sense in many instances, but dams are also damnable for several reasons: Dams alter habitats, threaten species, alter water quality, and generally do not remain viable for more than a century or two.

Michael Zuzel quotes Rebecca Wodder, president of the conservation group American Rivers: "We have blocked the flows, straightened the curves and hardened the banks of thousands of miles of waterways. By changing the most fundamental qualities of rivers—their natural shapes and flows—we've made it difficult for them to support life." Dams on many North American rivers cause native freshwater species—including several varieties of fish, mussels, crayfish, frogs, and snails—to go extinct as fast as some species living in tropical rainforests. Dams on U.S. rivers have transformed the once free-flowing river into a series of slack water ponds, thereby bringing salmon and steelhead to the brink of extinction.

The National Hydropower Association claims that hydropower dams produce pollution-free power while enhancing biodiversity and improving habitat. However, Anthony Ricciardi, a freshwater biologist at Canada's Dalhousie University, is unimpressed with this assertion. He declares that dams pose a major problem for the ecological health of rivers, and contribute to other problems. Amy Souers quotes him: "We also have to look at water quality, organic and chemical pollution and runoff from streets and yards...The invasion of exotic species—the zebra mussel for example—is also something that has to be addressed." Ricciardi also points out that if the current trend of river and wildlife destruction continues, then more species will be lost in the next century than during the past century.

About 83 percent of water used in the United States goes to producing electricity and irrigating crops.

## India

India's water issues are among the most severe on earth. With approximately 18 percent of the world's population, India has less than

4 percent of the earth's fresh water. Two-thirds of the Indian people lack access to clean water. More and more extreme measures are taken every day to access water, often just for basic needs. Selling water often produces more profit than growing a crop.

The Indian government promises 10 gallons per person per day, but it has failed to come close to that number in some areas. In many parts of India, inhabitants use as little as one-half to three-quarters of a gallon of water per day for drinking. The rest—2–9 gallons per person per day—is used for cooking, bathing, hygiene, and sanitation. Also, Indians use a surprising amount of water to produce food, particularly meat.

There are 23 million wells in India. As the population grows from today's 1.1 billion people to approximately 1.7 billion in 30 years, Indians will have to keep drilling. But as one digs or drills deeper, saltwater and other contaminants begin to seep into the shafts, rendering entire aquifers useless. Conflicts already rage throughout India over water, and farmer suicides are occurring by the thousands because of drought and water-related poverty and debt.

## China

China has a lot to worry about too. The Chinese character for "political order" is based on the symbol for "water." The clear meaning is "those who control the water, control the people." The mighty Yangtze, the backbone of the country's economy, is so polluted that by 2011 it may not be able to sustain marine life, much less human life. As if China's energy problems are not enough, the Chinese people grow about 70 percent of the nation's crops in a region having only 20 percent of its water. Also, there is growing competition for water among cities, industry, and all phases of agriculture. The water level beneath the North China Plain fell by 25 feet from 1991 to 1996.

## Africa

Africa's rivers support the economies, food production, and drinking water for their respective regions. The largest rivers and many other African rivers are polluted to catastrophic proportions. For example, only half the population in Nigeria has access to clean water, and many Africans must walk hours to get fresh water. A United Nations report predicts that water will be the major cause of conflict in Africa in the next 25 years. There is already conflict over water for irrigating and generating power. But what is the point? Both the winner and loser will still be running out of water.

### The Middle East

Water has historically been the most valuable resource in the Middle East. The politics of the region are closely linked to water and access to it. Muslim Turkey and Jewish Israel recently entered into a military alliance based in large part on Turkey sharing water with Israel. On a per capita basis Israel uses 4 times the amount of water used by people in Palestine, and this imbalance causes considerable tension.

### Central Asia

The once-majestic Aral Sea, one of the world's largest inland seas, used to support a large fertile area, but mismanagement has turned this area into a toxic wasteland rendering it unfit to supply the water needed in that region. The two rivers feeding this lake were diverted to grow cotton in the desert. From 1962 to 1994 the level of this once great sea fell by over 50 feet.

### Mexico

Mexico City, one of the largest cities in the world, is at risk of running out of fresh water. Lakes have been drained and forests cut down to accommodate the growing population. So much water has been pumped out from beneath the city that Mexico City is sinking. A water disaster is sure to come. Old pipes and gross mismanagement contribute much to the problem. For example, 40 percent of the city's water leaks back into the ground from leaky pipes, many cracked or displaced by the sinking ground; others fractured from age and neglect.

## Europe and Japan

On a per capita basis, Europeans and Japanese use about one-half the water that people in the United States use. Still, over half of European cities use ground water at unsustainable rates. About 5 million Spaniards experience chronic water shortages. The proposed solution is to redirect water from the Rhine River in France. I suspect this plan may upset some French citizens and others relying on the Rhine. Wouldn't a desalinization plant using renewable energy be a much better and safer alternative? Many Europeans and Japanese think so.

## Australia

Australia is the driest continent. Australians have tried to correct water problems by various innovative techniques. The Murry-Darling Basin produces over 70 percent of Australia's irrigated crops. However, many worry that overuse of irrigation water will make a large part of this area unusable for irrigation in 20 years. The availability of clean drinking water will become a problem at the same time. Another water-related problem is saltwater, which often moves in to replace the extracted freshwater. In some regions the water tables are rising, thereby pushing deadly quantities of saltwater to the surface. The saltwater destroys once-fertile farm land.

## WHAT CAN BE DONE?

Conservation and more recycling always make a lot of sense. One example was the redesign of the toilet, which decreased the water required per flush from 6 gallons to 1.6 gallons. This redesign saves approximately 5 billion gallons of water per year in the United States alone. Yet you can't recycle what you don't have in the first place. Residents of the United States, Europe, Hong Kong, Japan, and Australia have significantly decreased their per capita use of water over recent decades. In the United States, per capita use has decreased 25 percent since 1950, but the total population has grown 33 percent, resulting in a net increase

SOLUTIONS

in water use. Japan's use of water to produce $1 million worth of goods dropped from 13 million gallons in 1965 to approximately 3 million gallons in 1989. No matter, water seems destined to take many people to a disastrous end unless the world "produces" more fresh water.

The only solution that can make a significant difference is one that supplements nature's production of fresh, clean water—in this case Mother Nature will need a lot of help. *The world desperately needs more non-polluting renewable energy to desalinate saltwater or to clean up contaminated or dirty water from other sources.*

How much energy would be needed? Really not very much. There are several desalinization options. Membrane separation is the most efficient, and it requires about 12 kilowatt hours to produce 1000 gallons of fresh water. An energy cost of 7.5¢ per kilowatt hour (U.S.) would put the cost of fresh water at 90¢ per 1000 gallons, or less than one-tenth of a cent per gallon. Water in New York City costs about $3 per 1000 gallons, while water in London costs about $7 per 1000 gallons.

For perspective, it would take 30 trillion gallons of water per year to provide everyone on earth with 13 gallons of water per day, considered a practical minimum. At 7.5¢ per kilowatt hour, the energy to produce this much water today would cost approximately 30 billion dollars, about what the world spends on "defense" in 10 days. Renewable energy—wind, solar, and fast neutron reactors—would have to be used to avoid air and water pollution. These costs are easily manageable, and would represent only a 2 percent increase in the cost of electricity worldwide. These costs will likely fall with improved membrane technology, although the population will grow in the meantime. Notwithstanding the water issue, the world cannot endure unbridled population growth. If only people could receive a drop of education along with every sip of water.

Education, education, education. Consider this macabre thought.

> The world desperately needs more non-polluting renewable energy to desalinate saltwater or to clean up contaminated or dirty water from other sources.

250

As more people maintain better health with access to clean water, the population of the earth will increase even faster. One hopes, however, that the world will learn to manage some of the potentially devastating effects of population growth—and perhaps population growth itself—by differently distributing resources and providing basic human services, such as health care, education, and access to water, medicine, and electricity. Compared to a girl without a basic education, a girl who receives the equivalent of a sixth-grade education gives birth to far fewer children, enjoys a better quality of life, provides a better life for her children, and makes a bigger contribution to her community, family, and economy. Pretty good payback for such a small investment.

But we can *not* as easily manage disease, epidemics, resource depletion, mounting pollution, and the crime, violence, and despair that accompany these conditions. This is no way for people to live, no way for a planet to exist. And there is no need.

## BOTTOM LINE

Here are several specific recommendations:

▶ Limit amount of water per acre that can be used for irrigation. In turn, such limits will encourage water-conserving irrigation practices.

▶ Build desalinization plants worldwide to stave off the inevitable, grave consequences of the lack of fresh water—the single most essential substance of life.

▶ Where possible, tax the use of water to encourage conservation and pay for desalinization plants. The tax, of course, could be waved for those that can't afford to pay.

▶ Manage aquifer water to end depletion and begin replenishment.

Some of these suggestions are easy to put down on paper, but will be difficult to implement. Yet when a practice is unsustainable, it *must* be changed.

*Chapter 13*

# THE HYDROGEN ECONOMY

## NOT NOW AND MAYBE NEVER

In 1973, when writing my first paper on the energy crisis, I believed that fuel cells using hydrogen would become the ultimate alternative to fossil fuels and the internal combustion engine. I was dead wrong. Yet articles on hydrogen as a fuel would fill a library. It's way past time to get more realistic. Besides, hydrogen is not itself a fuel source; it is just an energy carrier, like electricity.

## HOPE and HYPE

As Stephen and Donna Leeb point out in their book, *The Oil Factor*, "Hydrogen is the holy grail of energy research." You take water, any water, from anywhere. Then you take sunlight to split water molecules ($H_2O$), thereby freeing up hydrogen ($H_2$). The hydrogen can then be burned, recombining it with oxygen to generate energy to run factories and power homes and cars. The by-product is water. Fresh, clean water. In the lingo of alternative energy, if you can crack water with sunlight, then you have a source of cheap, limitless, pollution-free energy and abundant fresh water.

"Hydrogen powered fuel cells promise to solve just about every energy problem on the horizon," writes David Stipp in an article called "The Coming Hydrogen Economy." In September 2000, T. Nejat Veziroglu, president of the International Association for Hydrogen Energy, proclaimed, "It is expected that the petroleum and natural gas production fueling this economic boom will peak around the years 2010 to 2020 and then start to decline. [JS: Many believe it has already peaked, but who's to argue?] Hydrogen is the logical next stage, because it is renewable, clean, and very efficient."

There is no doubt that hydrogen would be helpful in the effort to get the most direct benefit from the sun's potential as a fuel. Unfortunately, we have been unable to convert enough of the sun's energy that hits the surface of the earth into useful energy. Indeed, the conversion rate with current technology is a very low 2 percent. Even at much greater conversion efficiencies, a solar-based hydrogen economy is far-fetched.

Many believe hydrogen and the so-called hydrogen economy will be ready for prime time in the relatively near future. Some of my well-informed engineering colleagues believe hydrogen will rescue us from an impending economic meltdown caused by the depletion of fossil fuels. One legislator told me not to worry about energy, because hydrogen is the solution, and "the solution is just around the corner."

The world cannot afford such false and poorly informed statements. Such declarations are dangerous, and they mislead just about everyone, including world leaders, to believe that the solution to the world's energy problems is close at hand. Without a disciplined plan and a quantified timetable to get us there, these kinds of statements are just hot air. Don't believe vague, generic claims. No hydrogen economy is going to happen any time soon—and it may never happen.

## THE LOW DOWN

Many believe hydrogen—a colorless, tasteless, flammable, non-toxic gas—is an ideal fuel because it is abundant and combusts without creating pollution. Hydrogen is certainly abundant. It is the most plentiful element on earth, found in every living thing and all fossil fuels. However, hydrogen is not found in nature as a stand-alone substance, except in minute amounts—about 5 parts per million (ppm)—in the air we breathe. Rather, hydrogen is always in combination with other elements. For example, every water molecule contains a pair of hydrogen atoms. Thus, hydrogen must be isolated to create a fuel. Said simply, hydrogen is *not* an energy source; it is a *carrier*, and it must be produced by separating hydrogen from other elements, such as separating water into hydrogen and oxygen. This separation process, called *hydrolysis*, requires a lot of money, facilities, and energy. In short, the cost to produce hydrogen is simply out of economic reach any time in the foreseeable future.

Several additional problems plague a potential hydrogen economy. First, separating hydrogen from the elements it combines with in nature *always* uses more energy than is contained in the resulting hydrogen. If the hydrogen is derived from any source other than water, then the separation process produces greenhouse-gas emissions. Hydrogen derived from water also adds greenhouse gases, unless the electricity used to separate the hydrogen is generated from nuclear energy or some other renewable, non-polluting source of energy. But if you have the electricity, why convert it to chemical energy and then back again to electricity?

Second, while hydrogen can be burned directly in an internal combustion engine, it is most efficiently used in a fuel cell, which chemically recombines hydrogen and oxygen to produce electricity. This technology is very attractive because its efficiency is approximately double that of an internal combustion engine. However, existing fuel-cell technologies have a long way to go before they are ready for the market.

Third, the logistics of handling hydrogen as a gas or liquid are impractical. The element hydrogen is the lightest of all gases, 15 times lighter than air. One gallon of hydrogen weighs only 0.6 pounds. As a liquid, hydrogen is 14 times lighter than water. However, hydrogen cannot exist as a liquid at atmospheric pressure at a temperature higher than *minus* 423 degrees Fahrenheit. No matter how you look at it, hydrogen is extremely difficult to transport or handle as a very light gas or as a super-cold liquid.

Fourth, the energy content in a gallon of hydrogen is only 27 percent that of the energy contained in a gallon of gasoline. This means it takes approximately 3.7 gallons of hydrogen to have the same energy content as a gallon of gasoline. Here are the numbers:

- One gallon of hydrogen contains 31,000 BTU.
- One gallon of ethanol contains 76,000 BTU.
- One gallon of gasoline contains 114,000 to 125,000 BTU.

Other problems exist. Let's take a best-case situation. Let's say all of the many technical problems associated with hydrogen are solved. Then only the distribution problem remains. Still, the country would be many decades away from adopting hydrogen for large-scale production of electricity or use as a fuel for transportation. Experts estimate that the necessary infrastructure for distributing hydrogen—that is, an infrastructure for storing and distributing hydrogen that is comparable to the scope and capacity of the infrastructure in place today for automotive fuels—would cost at least 1 trillion dollars, and perhaps several trillion dollars.

The conclusion is clear: There will be no hydrogen economy any time soon. We are decades or generations away. Personally, I don't think hydrogen will ever be an important transportation fuel. Creation of the so-called hydrogen economy requires several major technical breakthroughs and many

> There will be no hydrogen economy any time soon. We are decades or generations away.

minor breakthroughs. Achieving such breakthroughs will be a very tough task, even with a lot of time and money. From experience I can tell you that if a business must rely on a technical breakthrough to succeed, then its chances for survival are slim. If it must rely on 2 or 3 major breakthroughs, then forget it. For example, even the production of inexpensive fuel cells has been a tough nut to crack, as costs have not yielded to heavily funded research. Fuel cells work, but the price needs to be *reduced by a factor of 10* to be of interest to manufacturers and the public. Achieving such an innovation or breakthrough will be a formidable task, especially given the lack of success to date, in spite of the money and effort devoted to this endeavor.

## A DEBATE

The great promise and possibilities of the so-called hydrogen economy have been debated for the last 30 years. Below is one of the latest debates, moderated by Mitch Jacoby, which appeared in *Chemical and Engineering News* in 2005. One of the debaters is Dr. Steve Chalk, a development manager for hydrogen and fuel-cell research at the U.S. Department of Energy

> The necessary infrastructure for distributing hydrogen would cost at least 1 trillion dollars, and perhaps several trillion dollars.

(DOE). In 2006 he was also the acting manager for the DOE's Solar Technology Program. Dr. Joseph Romm presents the opposing point of view. He is Principal Deputy Assistant Secretary for Energy Efficiency and Renewable Energy at the DOE. Both men have impressive credentials and backgrounds, and we can assume a great deal of experience with all aspects of energy production and deployment. Excerpts from the introduction and debate appear below, along with my comments in *italics*.

## Introduction to the Debate from the Editors of *Chemical and Engineering News*

In terms of atomic size and structure, hydrogen stands out as small and simple. Yet discussion of hydrogen's possible future role as a primary *energy* carrier sparks debates that are big and complicated.

For years, some scientists and policymakers have argued that replacing petroleum-based fuels with hydrogen in a so-called future hydrogen economy would be an ideal solution to numerous energy-related problems. For example, using hydrogen as the fuel in stationary and automobile-based fuel cells would cut down on pollution, they say, because fuel cells running on hydrogen generate electricity and water but produce virtually zero pollutants. Switching to hydrogen as the primary fuel also would lessen the growing U.S. dependence on foreign oil, proponents argue.

*No person would disagree with this, if you can get there.*

The case in favor of hydrogen has been made in academic circles for years. In 2003, President George W. Bush gave the idea of the hydrogen economy a shot in the arm and boosted public awareness by announcing the launch of a $1.2 billion hydrogen research initiative. Now [2005], as the price of crude oil exceeds $60 per barrel and prices at the gasoline pump approach $3 per gallon in some states, arguments in favor of hydrogen can be heard loud and clear.

*If we can't get it to work relatively soon, then it will deflect our attention from real solutions, and it will cost a lot of money and time. Since this debate appeared in print, the cost of crude oil has risen to $117 (U.S.) per barrel, gas prices have exceeded $3 per gallon, and prices are rising.*

But critics point out that the overwhelming majority of hydrogen prepared industrially—more than 90%—is made from fossil fuel sources such as natural gas via reforming processes that produce carbon dioxide and other greenhouse gases. And although hydrogen can be produced renewably—for example, by using wind- or solar-cell-generated electricity to electrolyze water—those processes are expensive.

In addition, critics contend that even if pollution and petroleum dependence are taken out of the equation, questions remain concerning the safety of hydrogen use by large numbers of motorists. Critics also stress that a storage and distribution infrastructure similar to the one in place today for automotive fuels will need to be designed, built, and tested before it can be used to supply a nation with hydrogen.

*At a likely cost of trillions of dollars. Luckily, we have better alternatives.*

Furthermore, technology for carrying an adequate supply of hydrogen compactly on board an automobile will need to be invented to provide motorists with an acceptable driving range between hydrogen fill-ups.

*Lots of money has been poured into this problem for decades, and I believe we are still very close to square one. If hydrogen gas were used, heavy high-pressure cylinders would be required. Not a good solution from the perspective of weight or efficiency. If hydrogen were used as a liquid, a vulnerable cryogenic vessel would be required—also not good. In the 1970s, under my direction, MVE Cryogenics built 20 hydrogen fuel tanks and gave them free to any organization doing research on hydrogen-fueled vehicles. We proved that an efficient liquid hydrogen tank could be built for an acceptable price, but we did not prove it could survive on the road.*

## Dr. Chalk: A Proponent

Business-as-usual approaches have not reversed and cannot reverse our increasing dependence on imported oil. Since the oil crisis of 1973, oil imports have grown from about 35% of total U.S. consumption to over 55%. This dependence is caused primarily by the transportation sector's growing demand and overwhelming reliance on petroleum. In fact, by 2025 imports are projected to be 68%, threatening both energy security and economic stability. The Hydrogen Fuel Initiative addresses our long-term dependence on imported oil while reducing emissions of pollutants and greenhouse gases. A recent National Academies [of Science] report supports this concept and concludes, "A transition to hydrogen as a major fuel in the next 50 years *could* fundamentally transform the U.S. energy system, creating opportunities to increase energy security through the use of a variety of domestic energy sources for hydrogen production while reducing environmental impacts, including atmospheric $CO_2$ emissions and pollutants."

*A transition to a hydrogen economy could fundamentally transform the national energy system, but if it takes 50 years, then it will be way too late. Moreover, inherent safety issues would still be a big problem, as would the cost.*

The Department of Energy's plan shows that it will take decades to fully realize the benefits of hydrogen.

*I haven't been able to find the plan Dr. Chalk mentions. Is it quantified? Costs? Milestones? Completion date?*

Therefore, DOE is continuing to develop high-efficiency technologies for near-term hybrid vehicles. The DOE FreedomCAR Program is investing over $90 million per year to make hybrid batteries, electronics, and materials more affordable. Hybrids are emerging in today's market and will provide the best approach for reducing petroleum consumption over the next 20 years. But efficiency alone will not enable our transportation sector, which accounts for two-thirds of our oil use, to eliminate its dependence on oil. Dramatically improved efficiency can slow the growth in transportation energy demand, but after a while the increase in population, the number of vehicles, and the miles of travel will cause oil demand to resume its upward trend. Therefore, fuel substitution must accompany fuel efficiency to achieve long-term energy security.

*Conservation should be a part of our energy strategy—Dr. Chalk is dead right about that. But conservation buys the world only 10 years or less to find and deploy other sources of energy. And Dr. Chalk is right about another point: "fuel substitution must accompany fuel efficiency," but we should replace gasoline with biofuels and electric motors, not with hydrogen.*

As a substitute for gasoline or diesel, hydrogen provides a long-term solution.

*The word "provides" implies a great leap of faith. Hydrogen provides promises and potential only, but definitely not a solution. Chalk's claim strikes me as a very dangerous, bet-the-farm assumption that can easily mislead everyone, especially leaders and policymakers.*

In partnership with auto and energy companies, ... the DOE program brings together leading scientists and engineers from hundreds of institutions—including university, industry, and government laboratories—to address the key challenges. These technical and economic challenges are formidable but not insurmountable:

Improving hydrogen storage energy density by a factor of three, allowing a vehicle range of greater than 300 miles.

Increasing fuel-cell durability five-fold and lowering cost from $200 per kW [kilowatt] (today's projected high-volume cost) to less than $50 per kW [kilowatt].

Reducing hydrogen cost by a factor of four to be competitive with gasoline.

DOE is also addressing hydrogen safety. Like other fuels, hydrogen can be used safely with appropriate engineering and handling. In fact, industry produces over 9 million tons of hydrogen annually and safely operates hundreds of miles of hydrogen pipelines. To ensure safe commercial use of hydrogen, DOE's program includes underlying safety research leading to new materials and components and new practices and building codes.

*C'mon. This is taking on the feel of a fairy tale. Hydrogen safety in the hands of millions of motorists is very different from industrial handling.*

Unlike previous alternative fuel vehicle programs, this initiative includes neither quotas nor sales targets; criteria for the commercialization decision are market driven. If successful, the Hydrogen Fuel Initiative will facilitate a 2015 decision on commercialization.

With subsequent investment in vehicle manufacturing and refueling infrastructure, hydrogen fuel-cell vehicles could enter the market in the 2020 time frame. Replacing the existing vehicle fleet takes time; therefore, significant energy and emissions benefits will occur after 2030. However, the DOE strategy of development-advanced hybrid vehicle technologies enables the country to begin the transition to hydrogen and fuel-cell vehicles while achieving petroleum savings and emissions reduction in the near term.

*Too little, too late.*

Transforming to a hydrogen-based transportation system is synergistic with reducing greenhouse gas emissions in the power sector. Because the total amount of energy required for the transportation sector rivals that for electricity generation, increased hydrogen demand can stimulate the expansion of carbon-free renewable and nuclear power. The U.S. is investing more than any other country in carbon capture and sequestration technologies, enabling virtually carbon-free hydrogen and electricity from America's abundant fossil resources such as coal.... Business-as-usual approaches haven't worked, and it is time to get serious about our over reliance on foreign oil.

*Sequestering carbon dioxide is expensive and potentially dangerous if large amounts of carbon dioxide were to escape from the sequestered site. Where should the site be? And why bother sequestering carbon dioxide when a better solution exists? Also, Dr. Chalk made a big leap from "investing" to "enabling"—kind of like confusing "is" and "may." All too often, investing does not produce expected results.*

*Let's see what Dr. Romm has to say.*

## Dr. Romm: A Critic

The time has come for action on global warming. The scientific consensus is strengthening that human-induced global warming will not be on the mild side, and may well be catastrophic if we do not quickly start cutting greenhouse gas emissions. Hydrogen cars have little chance of being a cost-effective strategy for reducing those emissions through 2035. They should be put on the back burner while we push fuel efficiency and hybrid vehicles now, followed quickly by hybrids that can be plugged into the electric grid.

British Prime Minister Tony Blair said in September 2004 that he believes climate change is the world's "greatest environmental challenge" and committed to "reduce our carbon dioxide emissions by 60% by 2050" to avoid catastrophic climate change. So we must move as fast as possible to zero-carbon sources of energy for our electricity and transportation fuel.

Yet even two well-known California hydrogen advocates, Joan Ogden and Daniel Sperling of the University of California [at] Davis, acknowledged in a 2004 *Issues in Science & Technology* article, "Hydrogen is neither the easiest nor the cheapest way to gain large near-and medium-term air pollution, greenhouse gas, or oil reduction benefits." So a focus on hydrogen represents a misdirection of resources away from strategies that can achieve far larger benefits for far less money for decades to come.

When will hydrogen fuel-cell cars be practical? As Bill Reinert, U.S. manager of Toyota's advanced technologies group said in January 2005, absent multiple technology breakthroughs, we won't see high-volume sales until 2030 or later. Reinert was asked when fuel-cell cars would replace gasoline-powered cars, and he replied, "If I told you 'never,' would you be upset?"

We need a major breakthrough in fuel-cell technology to bring down the cost by more than a factor of 10 while increasing durability and maintaining efficiency. And as a March 2004 report by the American Physical Society concluded, "A new material must be discovered" to make onboard hydrogen storage practical."

*Although Dr. Chalk states that the cost needs to be reduced by only a factor of 4, the consensus is closer to 10, as Dr. Romm states. Needing a "new material" is just one of many daunting breakthroughs required.*

Absent breakthroughs, hydrogen cars will remain inferior to the best clean cars available today, gasoline-electric hybrids such as the Toyota Prius, in virtually every re-spect—cost, range, annual fueling bill, convenience, safety—and in providing cost-effective reductions of greenhouse gas emissions and oil consumption.

*For transportation, plug-ins and all-electric cars are the final solution—and within reach today. Why must we have automotive fuel cells? Am I missing something? Why use electricity to make hydrogen with all its headaches, when we can use the electricity directly? Simplicity is a beautiful thing.*

Don't get me wrong. I favor keeping the hydrogen option open. I helped oversee the Energy Department's program for clean energy and alternative fuels, including hydrogen, for much of the 1990s, during which time we increased funding for hydrogen ten-fold. But DOE is cutting the budget for efficiency and renewables—technologies that can reduce greenhouse gas emissions cost-effectively to-day—to fund the hydrogen program, which cannot do so anytime soon. This is a mistake.

Moreover, a 2004 report from the European Union's Joint Research Centre found that hydrogen cars would likely increase greenhouse gas emissions. Hydrogen is not a primary fuel, like oil, for which we can drill. It's bound up tightly in molecules of water or hydrocarbons such as natural gas. A great deal of energy must be used to un-bind it. Making that energy causes pollution.

Delivering pollution-free hydrogen to a car is expen-sive, likely costing the equivalent of $6.00 per gallon of gasoline (untaxed) or more for a long time. So we get a bait and switch, with politicians promising renewable hy-drogen but then subsidizing polluting hydrogen filling stations. More than 95 percent of U.S. hydrogen is made from fossil fuels and, as a prestigious National Academy of Sciences (NAS) panel concluded in 2004, "It is high-ly likely that fossil fuels will be the principal sources of hydrogen for several decades."

Furthermore, using renewables to make hydrogen is simply bad policy, even if prices drop sharply. Renewable electricity can achieve far greater pollution reduction by directly displacing coal—or even natural gas—in the power sector. And those savings can be achieved without a massive investment in the hydrogen infrastructure.

So we are several decades from a time when serious investments in hydrogen cars or infrastructure make sense environmentally. While we wait, we must push fuel efficiency and advanced hybrid vehicles to address the urgent problems of global warming and oil imports. We should promote ethanol from sources other than corn as a gasoline blend and begin deploying hybrids that can be plugged into the grid and can run four times as far on a kilowatt-hour of renewable electricity as fuel-cell vehicles.

Hybrids now; hydrogen much, much later.

Dr. Chalk rebuts, and Dr. Romm responds in turn, but the fundamental arguments don't change. The *dream* of hydrogen and the hydrogen economy will remain just that—a dream.

I don't envision a hydrogen economy in the next 50 years, or even in this century, no matter how much money is thrown at the problems. Besides, why put all that money into such an elusive technology when *we already have a better, more immediate solution*, a solution that is more compatible with traditional energy distribution?

## AUTO INDUSTRY

Has anyone in the automotive industry presented the case for hydrogen cars? Has any source detailed how much energy would be saved and what reductions in emissions could occur if the hydrogen economy became a reality? *Car and Driver* did in its October 2006 issue. Patrick

Bedard presents the case. Bedard references an expert, Dr. Donald Anthrop, Ph.D., who did the necessary calculations. Anthrop is professor emeritus of environmental studies at San Jose State University. Here are some excerpts. I paraphrase.

- The overall efficiency from hydrogen production to useful energy to propel the car, including either compressing the gas or liquefying it, is only 12 percent. Therefore, hydrogen production for transportation use consumes roughly twice as much energy as the use of gasoline.

- If the electricity to produce the hydrogen came from the current mix of electricity-producing energy sources—coal, natural gas, nuclear, hydro, wind, and solar—then the carbon emissions would double compared to the use of gasoline.

- Hydrogen can also be produced by steam reforming natural gas. This process is only 30 percent efficient. Consequently, burning the natural gas directly makes more sense from the perspective of energy consumption and emissions.

Mr. Bedard concludes the article with the following:

> Presumably BMW knows all of this, yet it has been thumping the tub for hydrogen since the 1970s. Along with hundreds of other attendees, I attended BMW's hydrogen hootenanny at Paramount Pictures in 2001. Mostly, it amounted to a day of preening before California's greenies…I summed up the science of this column, in writing, and passed it up through BMW's official channels, along with the obvious questions: Where will the necessary quads and quads of energy come from for hydrogen cars? [Note: a quad is 293 billion kilowatt hours.] That was nearly two years ago. BMW has not answered.

No answer, of course, is the answer.

## BOTTOM LINE

▶ Hydrogen has often been called "the fuel of the future," and it probably will always remain so.

▶ We cannot count on the so-called hydrogen economy to help us anytime soon—period. Let's not let the possibility of a hydrogen economy distract us from solving the world's daunting energy challenges with proposals founded in real possibilities.

# Part Three

## A NEW DAWN

## Chapter 14

# TRANSPORTATION

Cars and trucks are not merely transportation vehicles. They are symbols of who we are as individuals and as a society. Some vehicles putter along with practicality and convenience. Others roar with speed and flash. Our vehicles reflect some of our values, including some values that are misplaced. Although many people die in their autos, fossil-fueled vehicles may be the death of us all. On the other hand, new technologies may lead us to a clean transportation future. Let's see.

The current transportation system has lots of problems. The convenience and necessity of travel by car has given us heavy-duty pollution, urban sprawl, smog, and even road rage. Complaining about traffic is a national pastime ... and for good reason. Americans take 411 billion trips and drive about 4 trillion miles annually. The average length of a trip is 10 miles. Traffic congestion is getting worse in the United States and throughout the world. Traffic delays in the United States cost $65 billion a year according to the Texas

> Americans take 411 billion trips and drive about 4 trillion miles annually. The average length of a trip is 10 miles.

Transportation Institute, and this cost is rising. In the United States the annual delay per driver is almost 50 hours per year. If the pollution doesn't affect your health, the stress might.

Traffic jams waste more than 2.3 billion gallons of fuel every year. This wasted fuel puts 20 million tons of $CO_2$ and other pollutants into the air. At that rate you can see how we throw stupendous amounts of $CO_2$ and other foul stuff into the atmosphere even when idling.

We can be more precise. In the United States there are approximately 8 cars for every 10 people. That calculates to approximately 250 million cars consuming approximately 5 billion barrels of oil per year. That's 1.5 billion tons of $CO_2$ per year. And that's just cars and light trucks in the United States. Forget other countries, large trucks, buses, planes, trains, ships, bulldozers, golf carts, riding mowers, snowmobiles, and so on. The $CO_2$ has no place to go except into the air we breathe and into the world's oceans—damaging plant, animal, human, and aquatic life.

> Each vehicle creeping and crawling to its destination spews 17 pounds of $CO_2$ and a quantity of other pollutants into the atmosphere for every single gallon of gasoline burned.

To make matters worse, the global market for new cars and small trucks is growing rapidly. According to *Economist* magazine (February 24, 2007), more vehicles will be produced in the next 20 years than during the whole of the twentieth century. Some 180 new factories, each producing 300,000 vehicles, are set to double global annual production to 110 million units a year—a terrifying thought. Most cars will be basic and designed to sell for $3000 to $13,000. Some will be more motorcycle than car. If the internal combustion engine remains the main power plant, then world oil reserves will *not be able* to sustain this market—period. See Depletion Chart (Figure 1.1).

As demand for gasoline rises and supplies dry up, gasoline prices will skyrocket. Expect prices in the range of $5–10 per gallon in the relatively near future. We will also see gasoline rationing. It is inevitable. The sooner the better. Again see the Depletion Chart, which shows fossil fuels falling to zero, but we all know huge problems with supply and cost will emerge well before then.

Before turning to transportation options for the future, let's look to the past for lessons learned and to the present for details of current transportation problems.

## CHALLENGES

Can we learn anything from the past about our present challenges? Few people recognized any pressing transportation problems 100 years ago. After all, horses and horse-drawn vehicles had been the primary forms of overland travel for many centuries. What horses could not do, railroads and steamboats could. As the human population grew, the need for horses grew. The more horses, the more horse "pollution." In the early 1900s, even small cities were burdened with—or buried under—horse manure. In Chicago, where over 10,000 dead horses had to be hauled away every year, tens of thousands of pounds of horse manure piled up every day. Imagine the manure dust, the stench, the flies, the pathogens, and the illness. The main form of transportation created enormous environmental and health problems. To the rescue came the automobile. Gasoline at the time was less than 20¢ per gallon in today's dollars—a tremendous bargain compared with the cost of the feed, tack, and stables required to maintain the horse-centered transportation system, and it was cleaner.

The world again faces a huge problem with too much of a good thing. The cost of car fuel (feed) rose almost 40 percent in 2007, with no relief in sight. Now add the climbing environmental costs that humanity

> In 2007 sales of SUVs rose over 20 percent and sales of gas guzzlers rose 15 percent.

cannot afford. Yet, it is difficult to change the driving habits of America and the world. People "love" their cars as they love their pets and used to love their horses. To change driving habits and transportation preferences is often not about making rational choices. Rationality and reason often do not apply when people think about cars. Is love rational? For example, in 2007 the sales of sport-utility vehicles rose over 20 percent. U.S. sales of huge, luxury gas guzzlers are up 15 percent. Sales will remain strong, so exhaust pollution will continue to foul our skies and darken our prospects. This is difficult to understand with all the rhetoric about global warming and pollutants. One wonders if people are listening, are desensitized, or just don't care. Instead of stopping the march to the disastrous end some predict, we in the United States continue our wasteful, consumptive ways—*that's not rational.*

What is rational is to end government efforts to keep the price of fuels artificially low. I don't believe America's car-centered culture and its ho-hum attitude will see much change until drivers are staggered by the price of fuel. No pain, no gain? But then it will be too late. Can you imagine $200 or $300 per tank full? It is coming.

What do people pay elsewhere? In November 2007 gas sold for $7.76 per gallon in Norway, $7.00 in Germany, and $6.50 in France. That's why you see so many smaller and fun-to-drive cars in Europe. Europeans don't like our gas-guzzling habits. On a per capita basis, they use half the fuel we use. Europe has much better public transportation and higher population densities, which partially explains their need for less automotive fuel.

> More vehicles will be produced in the next 20 years than during the whole of the twentieth century.

### Reserves and Consumption

Let's start with a few facts. I note these facts elsewhere, but they bear repeating. In 1800, decades before the beginning of the Oil and Gasoline Era—whether it started in West Virginia in the 1820s or at the Drake well in Pennsylvania in 1859—the world was blessed with approximately 2 trillion barrels of conventional oil that took Mother Nature many hundreds of millions of years to produce. We've used up about half.

Let's imagine the world could produce another trillion barrels or more from unconventional sources such as the oil from the oil sands in Canada, heavy oils in Venezuela, or the oil shale in the United States. However, these sources will require a great deal of energy to extract. Reserve estimates of these vast unconventional sources vary greatly. Much is there, but the oil actually recoverable with present technology stands at approximately 315 billion barrels from oil sands, 35 billion barrels from heavy oils, and 580 billion barrels from oil shale. The cost to produce oil from these unconventional sources is $25–40 per barrel. While I cannot quantify the environmental damage of the techniques for extracting and processing these unconventional sources, the damage is reported to be very significant—but repairable.

Even if we extract and consume these unconventional resources, how long will they last? Let's look at some details and make a few reasonable assumptions.

*Baseline Assumptions About Population and Oil Use*

1.  The United States—about 5 percent of the global population—consumes approximately 7.5 billion barrels per year. The remaining 95 percent of the world uses approximately 22.5 billion barrels per year, making total global oil consumption about 30 billion barrels per year.

2.  Predicted population growth rates are all over the map (see Chapter Three on population). The consensus holds that the world's population is growing at approximately 1.25 percent per year. The U.S. population grew 1.24 percent from 1996 (265 million) to 2005 (296 million). Let's use 1.25 percent for global population growth and 1 percent for U.S. population growth.

3.  Assume consumption of oil grows mainly in proportion to population growth with some additional consumption from economic growth in China and India.

Combining current rates of population growth with current rates of oil use, we'll run out of oil in 28 years. What if we *double* our total oil? At current consumption rates, if we double our total oil with 1.1 trillion barrels from unconventional sources, then the oil would last 74 years. That's just to 2082. Yet again, we'll actually run out much sooner, because if we combine population growth, some economic growth, and current consumption rates, then total oil reserves will last less than 60 years. That gets us to 2068.

The bottom line is inescapable: We don't have much time. The oil is running out. And look, China is just beginning its love affair with the car.

## Step on the Gas Pedal and Hand Me a Gas Mask

This is what *Economist* (June 4, 2005) had to say about the growing auto culture in China. Below are some excerpts from under the heading "China is not yet an auto-culture in the mould of the United States. But it may only be a matter of time."

> "China has begun to enter the age of mass car consumption. This is a great and historic advance." So proclaimed the state-run news agency, Zinhau, last year. Environmentalists may feel a twinge of fear at this burgeoning romance with motoring. But a rapid social and economic transformation is under way in urban China, and the car is steering it.
>
> China's rapidly growing dependence on imported oil—one-third of it now for car use—is causing deep anxieties about the country's energy security. Petrol consumption will no doubt be curbed by taxes, better technologies or the use of alternative fuels.
>
> China has fallen in love with cars; and despite government efforts to cool the passion down, it burns as hot (and as noxiously) as ever.

Has anybody asked where the needed oil and gasoline will come from? China and India are lucky—they can go directly to plug-in hybrids and all-electric vehicles. Will they? If they do, they may cause all automobile manufacturers to enter into the new age of motoring, and that would be a virtue.

Is the world doomed to inhale exhaust fumes until the engines quiet and the lights fade? No. We are not doomed, but we need to take our heads out of the proverbial sand. Each time someone slips behind the wheel, we all slide a bit further behind the 8-ball. The world needs to act fast, to move rapidly to hybrid plug-ins and all-electric vehicles. We may already be too late to avoid an economic meltdown. Also, unless we act fast we may lose the economic and political ability to take the necessary steps to deploy the available solutions. However, to act quickly we need the world's leaders to understand the stakes and to act decisively. Just talking about "change" doesn't change anything.

For the moment let me draw your attention to the cars you drive, the firms that produce the vehicles, and the politicians who deceive you about their performance. Prepare to get steaming mad.

## EVEN POLITICAL FAVORS CANNOT CURE INCOMPETENCE

General Motors was the greatest American company of my youth. Ford was one of the greatest companies in U.S. history. A prominent saying was "As GM goes, so goes the country." Ford used to advertise that "Quality is Job 1." Still true? Today, GM is an automotive and corporate also-ran. Ford operates in the fading glow of its former glory. For decades, U.S. automakers simply have not been building world-class cars. Many observers think GM or Ford is headed for bankruptcy.

### Futility and Incompetence: GM, Ford, and Chrysler

GM loses money like a leaky bucket loses water. Ford hasn't produced a quality, world-class product in decades. U.S. designs haven't turned any heads for years. Where are the top-flight engineers, designers, and the stylists? American cars often look like they were designed by tired and bored executives, styled by accountants, and built by distracted workers. U.S. cars are no longer deft and wonderfully engineered triumphs. Instead, they are daft and woefully embarrassing trifles. Rarely does a GM product make it onto a Top Ten list for overall quality, but many are permanent fixtures in the Bottom Ten. The executives and professional staff at GM, Ford, and Chrysler should be required to read *Consumer Reports* and *Car and Driver*. Their customers do.

Perhaps GM and Ford officials were too busy building poorly designed and poorly performing vehicles to notice the big changes in energy, environment, and public dissatisfaction. In the face of our energy problems and soldiers fighting in oil-soaked lands, GM gave us the monster Humvee, apparently taking styling tips from the U.S. military. Toyota, Honda, and others gave us high-efficiency hybrids. There are no plug-in hybrids yet, but they are coming. The formerly great companies of GM, Ford, and Chrysler—with thousands of engineers, handsomely paid executives, and what are supposed to be top-notch Boards of Directors—are not responding to the country's energy woes or transportation tastes, and the companies seem not to know what to do about these troubles. If they would just look around, they would see an incredible opportunity and a possible rebirth. Fortunately, at last I see some positive stirrings at the so-called Big Three auto companies—GM, Ford, and Chrysler.

In my opinion, due to gross ineptitude GM has become one of the greatest *immobilized* forces on earth. The Chevy name is now synonymous with high maintenance and poor performance. GM vehicles—along with many Ford and Chrysler products—consistently come out at or near the bottom in competitive comparison tests (see Figure 14.1).

## Figure 14.1. Automobiles and Light Trucks Recommended by *Consumer Reports* in April 2007

| Company | Number of Vehicles Reviewed | Number Recommended | Percent Recommended |
|---|---|---|---|
| Honda | 15 | 14 | 93 |
| Toyota | 27 | 20 | 74 |
| Ford | 22 | 8 | 36 |
| GM | 44 | 13 | 30 |
| Chrysler | 22 | 4 | 15 |
| Mercedes Benz | 11 | 0 | 0 |

In the meantime our poorly run automakers are getting their technological, performance, and production butts severely kicked by foreign competitors.

This mounting futility, incompetence, and inefficiency will be difficult to overcome for the Big Three, even though current models look better, are more reliable, and are more efficient than models in the recent past. These legacies will be especially difficult to overcome because U.S. automakers are no longer in the habit of innovating and improving. In fairness, I like the look of Chevy's new hybrid Malibu. I rented one on a recent trip, and I must admit I loved it. It looks world-class in every respect. Only time will tell if its performance matches its looks.

GM lost market share because its executives didn't think the company would have to improve to survive. Instead, Big Three officials travel to Washington, D.C., to lobby for special treatment, favorable laws, and protection. Big Three executives were confident that Washington officials would protect them and their honored place in American history, industry, and economy, just as Washington officials had done for them in the past and just as they did to protect Chrysler from bankruptcy in the late-1970s and early-1980s. Even when they buy special favors in Washington, D.C., the Big Three U.S.

> U.S. automakers are no longer in the habit of innovating and improving.

automakers simply cannot compete successfully with other car manufacturers. People just don't trust GM, Ford, and Chrysler anymore, and they don't have to. They just buy Toyotas and Hondas, the world standard for efficient, world-class vehicles. Toyota's Lexus, Honda's Acura, and Nissan's Infinity further improve the package by adding terrific service and customer care to the mix.

### A Ford Man Defects to Toyota

Toyota, the undisputed leader in the worldwide car industry, has nearly displaced GM as the Number One builder of cars in the world. How? A former Ford executive, Jim Press, tells the story. Press, who was in charge of Toyota's sales in North America, left Ford in frustration 40 years ago because he did not think Ford handled customer relations properly. The *Economist* (January 29, 2005) quotes Mr. Press, "The Toyota culture is inside all of us. Toyota is a customer's company." He says, "Mrs. Jones is our customer; she is my boss. Everything is done to make Mrs. Jones' life better. We all work for Mrs. Jones." Even this simple concept has eluded U.S. automakers. Mr. Press has recently left Toyota to try to bring Toyota's magic to Chrysler.

> People just don't trust GM, Ford, and Chrysler anymore, and they don't have to.

### Advice for U.S. Automakers: Innovate or Wave Goodbye

GM, Ford, and Chrysler have similar problems, opportunities, and challenges. Each is in dire straits, but each is also lucky because profound changes in the auto industry and in society give each company new unprecedented opportunities, if—*if*—the companies adapt thoughtfully and *quickly*. Change brings opportunity.

Quit wringing your corporate hands and expending energy on bureaucratic trivia. Get back to basics. Your battlefield is lined with engineering, design, and production trenches. Your warriors are engineers, designers, stylists, and factory workers. Mobilize your vast engineering assets to become world-class developers, designers, and builders of plug-in hybrids and all-electric cars. Challenge your managers, designers, engineers, and factory workers to build the most effective imaginable versions of these automobiles. Stop lamenting the past, trim your staffs, bring management and labor together to work toward common goals, and begin building the cars of the future. An exciting future awaits if you find the courage to grab and shape it. In short, return pride to your company and employees.

Let's take a lesson from Chrysler's example of an old maxim: Necessity is the mother of invention. In the late-1970s and early-1980s the Chrysler Corporation teetered on the brink of bankruptcy. Congress agreed to back large loans to Chrysler, and the company had a new lease on life. The company had to turn around quickly, so it had incentive to try new ideas. Hal Sperlich, a young executive with the firm, and a small group of designers and engineers broke all the rules and came up with a company savior—the minivan. It was an incredibly successful and timely idea. What will be the Big Three's new "minivan," their breakthrough product, their legacy? Without great products, all other corporate activities will ultimately fail.

Gimmicks won't work. The public and commentators view the long warranties that GM and Ford recently offered as a desperate move previously reserved only for new entrants into the U.S. auto market. Think Hyundai and Suzuki. These long warranties are just another short-term fix that creates future liabilities.

Complaining won't work either. Automakers should stop going to Washington to whine and beg legislators not to mandate fuel-efficiency increases. Where is your once-honored leadership? As people say, "lead, follow, or get out of the way." Those people forget to mention that if you don't, you may get run over by the competition.

### I'm Too Slow to Get Out of the Way, So I'll Follow as Well as I Can

A GM executive, Chris Preuss, described GM's plans for plug-in hybrid vehicles as watching and waiting for "viable battery breakthroughs." Greg Gordon, in the *Minneapolis Star Tribune*, quotes

Preuss: "Right now we're not seeing a specific approach that would meet most of the expectations of consumers," he said, but if such technology becomes available, "there would be a fierce race to get that technology first." Does this statement sound like the plans of a leading, forward-looking company? Or the wishful-thinking of a follower hoping to get lucky? Waiting for someone else to create your good luck is not a sound plan. Why don't GM, Ford, or Chrysler develop or create their own future by creating the technologies they need? They sound resigned to wherever business as usual takes them. I can't imagine such a comment coming from what was once the world's top car company.

## THE PAST

### Fuel Efficiency

Americans, you are the victims of corporations that pay off politicians in order to secure favorable treatment to the detriment of the U.S. public. One of the most dishonest and dishonorable programs arising from such deals is called the Corporate Average Fuel Economy measurement. People call the program CAFE. We'll call it what it is: political acquiescence to a powerful special interest group.

### What Is CAFE?

Corporate Average Fuel Economy (CAFE) is the sales-weighted average fuel economy, expressed in miles per gallon (mpg), for any given model year of a manufacturer's fleet of passenger cars or light trucks. The Environmental Protection Agency (EPA) defines *fuel economy* as the average mileage traveled by an automobile per gallon of gasoline consumed (or the consumption of an equivalent amount of other fuel), as measured in accordance with the EPA's testing and evaluation protocol.

That sounds pretty wordy and technical, however the idea is pretty clear. What's the point of passing legislation to encourage car manufacturers to produce efficient vehicles and to make the efficiencies known to the public if the testing and reporting of the efficiencies is inaccurate and misleading?

In the midst of the oil crisis of the 1970s, Congress set mandatory fuel-economy standards for passenger vehicles in future years. The goal was to double 1974's passenger-car-fuel-economy average by 1985—that is, to raise the average fleet fuel consumption to 27.5 mpg. Congress set fuel economy standards for the intervening years:

18 mpg for model year 1978

19 mpg for model year 1979

20 mpg for model year 1980

27.5 mpg for model year 1985 and thereafter

Consumers Union, an organization that takes no money from manufacturers or anybody else, including advertisers, ran an exposé on this unholy alliance of car manufacturers and the EPA in the October 2005 issue of *Consumer Reports*, its consumer-advocacy and product-testing magazine. *Consumer Reports* declares that the EPA tests misrepresent the miles per gallon for cars and trucks 90 percent of the time and always in favor of the car manufacturers. *Ninety percent of the time.* The magazine is not the only one calling for more honest testing. Other articles in leading auto magazines also complain about the same thing.

Mr. Csaba Csere, a well-known car commentator, in the December 2005 issue of *Car and Driver* wrote,

> General Motors touts it has 19 cars that get more than 30 mpg. Toyota makes similar ad claims.
>
> And their claims are true, at least according to the EPA's highway fuel economy test. But there's not a snowball's chance that any of these vehicles will match their EPA-certified mpg when you drive them on a U.S. highway. In fact based on the fuel economy we get from *Car and Driver* test cars, most vehicles barely matched their EPA *city* fuel economy numbers while being driven on the *highway*.

## CAFE Penalties and More Confusion

Why is the U.S. government so unscrupulous with the mileage numbers and so cavalier with the American people? *Consumer Reports* speculates that U.S. government officials have "paid special favors to their big political campaign contributors."

The result? The U.S. government and special interest groups cost American drivers about 75¢ more for every gallon of gas they buy. This cost burden gets worse as gasoline prices rise. Today, the price of gasoline hovers near $3.50 per gallon. See Figure 14.2.

### Figure 14.2. Inaccurate MPG Claims for City Driving.

| Vehicle Type | Make and Model | City MPG | | |
| --- | --- | --- | --- | --- |
| | | EPA mpg | Consumer Reports mpg | EPA error |
| Small SUV | Jeep Liberty Diesel Ltd. 4WD | 22 | 11 | 50 |
| Hybrid | Honda Civic Sedan | 48 | 26 | 46 |
| Large Sedan | Chrysler 300 C | 17 | 10 | 41 |
| Midsized SUV | Chevrolet TrailBlazer EXT LT 4WD | 15 | 9 | 40 |
| Minivan | Honda Odyssey EX | 20 | 12 | 40 |
| Luxury Sedan | BMW 7 Series 745Li | 18 | 11 | 39 |
| Pickup | Dodge Ram 1500 SLT crew cab 4WD | 13 | 8 | 38 |
| Family Sedan | Oldsmobile Alero GL | 21 | 13 | 38 |
| Large SUV | Dodge Durango Limited 4WD | 13 | 8 | 38 |
| Small Sedan | Ford Focus ZX4 SES | 26 | 17 | 35 |

Source: *Consumer Reports*, October 2005.

### How You're Shortchanged

Inaccurate mileage figures can cause real pain in the pocketbook over the 5 years you're likely to own the vehicle. The extra fuel cost depends on the actual versus tested efficiency and how much of your driving is in the city. Moreover it varies from model to model: At least $3200 more for a BMW 7-series, an extra $3700 for a Chrysler 300C, and $3000 more for a Honda Odyssey EX. That assumes driving 12,000 miles per year, gas at $3 per gallon, and half of the annual driving done in the city. Even worse is the fuel we are wasting and the added pollution we put into the atmosphere—not to mention the balance of payments problems and U.S. troops in Iraq. *This is a big deal. A very big deal.*

*Who Benefits and Who Doesn't?*

**The winners:**

◆ Automakers, which avoid paying costly penalties.

◆ Government, which claims to do its jobs while keeping Big Oil happy.

◆ Oil companies, which sell a lot more product.

**The losers:**

◆ Consumers, who don't get what they pay for.

◆ Health, which is assaulted and compromised by pollution.

◆ The Environment, which is fouled and poisoned by pollution.

◆ The Economy, which takes a double hit—the balance of payments and the value of the dollar.

> The EPA tests misrepresent the miles per gallon for cars and trucks 90 percent of the time and always in favor of the car manufacturers.

By now you should be disgusted. Our government officials, in collusion with their automobile buddies, are totally frustrating the spirit of the law aimed at cutting gasoline consumption and lowering emissions.

And why did the average vehicle in model year 2006 weigh 500 pounds more than the average vehicle weighed in 1996? We are headed in the wrong direction just when world oil reserves are scarce and auto-created pollution is such a problem. Either automakers don't think global warming is a problem, or they don't care, or ignoring the problems is more profitable.

The automobile industry makes a lot of money on the hulking, behemoth vehicles they sell, and it repeatedly tells the public the monsters are safer. While larger vehicles fare better in a collision with smaller vehicles, more and more data show that monster-mobiles may not be safer at all. Why not? Because they have a strong tendency to roll over—the bigger they come, the harder they fall. In fact, according to a study from the *Journal of Pediatrics*, SUV's are 4 times more likely to roll over than a car. Also, the size and weight make them poor at avoiding accidents. If everybody drove smaller and lighter vehicles, this problem wouldn't exist.

## No Silver Bullet or Magic Remedy

If all vehicles today were as efficient as the EPA-specified CAFE number, then the nation would save approximately 40 billion gallons of gasoline per year—every year. Compare this potential savings of 40 billion gallons of gasoline per year to the promise advertised by British Petroleum to save 25 million gallons of oil—a drop in the ocean—by the year 2015 through various innovations and efficiencies. Now there's a fossil-foolish promise. Also, if all vehicles today were as efficient as the EPA-specified CAFE number, there would be 350 million fewer tons of $CO_2$ spewed into the atmosphere every year. It takes about 9 billion trees to use up that amount of $CO_2$ through photosynthesis. (Note: In December 2007, the U.S. Congress revised the CAFE standards. Full compliance is not necessary until 2020.)

## THE PRESENT

The competitive battles of the present and the future will be won by the companies that bring products to market that are most responsive to the world's energy realities. The winning vehicles will be plug-in hybrids and all-electric cars in the most attractive packages. Unsurprisingly, Toyota and Honda lead the march toward innovative development of hybrid vehicles. Small, forward-thinking companies are also introducing important

Either automakers don't think global warming is a problem, or they don't care, or ignoring the problems is more profitable.

innovations. All-electric cars are the ultimate goal, because the fuel costs for operating all-electric cars would be 6–10 times less than the costs for operating internal combustion engines, assuming a comparable "fuel" tax. Such vehicles will make a tremendous contribution to society and our environment. Most importantly, they will play a large role in our march to energy independence. By eliminating internal combustion engines, we can kiss oil shortages and their attendant pollution goodbye.

Of course, there are issues with hybrids, hybrid plug-ins, and all-electric cars, but they are already very functional, and all of the issues will be resolved with time. Toyota's Prius, for example, is far from perfect, but it is headed in the right direction and will improve. Cost is one issue. The sticker price for a hybrid is more than for a small, gasoline-powered car. However, operating costs for a hybrid, for example, should reclaim all or part of their initial cost disadvantage. Progress is already being made on alleged problems with the materials used to make batteries and extend battery life. Cold-weather driving is another limitation that affects some batteries more than others, and here again progress is being made. My Prius loses about 10 percent of its efficiency in our Minnesota winters. The makers of Phoenix and Tesla all-electric cars report that their vehicles have integrated heating systems. Such heating permits these vehicles to operate down to minus 40 degrees Fahrenheit without sacrificing much performance, according to Bob Goebel of Altairnano, battery supplier to Phoenix.

Patrick Bedard in *Car and Driver* confirms that GM is nowhere near the cutting edge of regular hybrids. Here are excerpts from the article:

> After its first cautious step toward hybrids in 2004 with full-size pickups, GM now takes a second [step]. The game plan for the Saturn Vue Green Line is less

All-electric cars are the ultimate goal, because the fuel costs for operating all-electric cars would be 6-10 times less than the costs for operating internal combustion engines, assuming a comparable "fuel" tax.

hybrid for less money. You get three shiny 'HYBRID' badges and a weak engineering effort for $22,995.

EPA mileage improves to 27 mpg in the city and 32 on the highway, from 22 and 27 for the 2.2-liter nonhybrid Vue...

Cheap and joyless, there's not even a mileage computer for keeping score of your adventures in thrift. Then again, there's little thrift here to bother with.

But perhaps *Car and Driver* exaggerates. A study released on June 29, 2006, reveals humbling conclusions for U.S. automakers, as reported by the Bloomberg financial news agency. "Toyota and Honda have done the best job building energy-efficient vehicles, while General Motors has done the worst." What do consumers think? "Seventy-seven percent of [American] consumers believe that either Toyota or Honda lead in developing vehicles such as hybrid gasoline-electric vehicles. GM is viewed as having the poorest record in developing alternative-fuel vehicles." How will GM counter this lousy public perception? "To compete with hybrids sold by Toyota and Honda, GM plans to introduce an SUV with an engine that runs on gasoline and electric power in the 2007 model year." Guess what? That vehicle was the Saturn Vue, described immediately above by *Car and Driver*. Perhaps Senator Pete Domenici, a Republican from New Mexico and a former Chairman of the Senate Appropriations Committee, had good reason for declaring, "Gee, those Neanderthals in Detroit just don't get it. We've got to save them from themselves."

So, the nagging concern about whether the world can find an alternative to the polluting internal combustion engine is resolved. New battery technology is the final enabler that makes hybrid plug-ins and all-electric cars a viable alternative to the internal combustion engine. GM, Ford, and Chrysler should get into high gear and mobilize their vast resources to develop world-class versions of such vehicles. There are reports that their new products are getting much better. We wish them good luck. Perhaps they will re-establish their automotive leadership.

So far we have discussed the past and the present: EPA promises to become more realistic. Higher, much higher standards are needed until we finally retire the internal combustion engine. In the meantime auto mileage stickers will be more accurate as of January 1, 2008. Very good news for the consumer.

I personally believe all companies, including GM, Ford, and Chrysler, will be introducing wonderful new products in the near future. Why? Because they must—their future depends on it.

## FUTURE TRANSPORTATION

Cars of the future *must* be cost efficient, fuel efficient, and carbon efficient. We can drive hybrids now, but in the future we must drive hybrid plug-ins and all-electric cars. Since hybrid plug-ins and all-electric cars are so efficient, much less fuel will be needed to generate the electricity needed to charge the cars' batteries than we use now by our transportation fleet burning gasoline. In effect, the use of hybrid plug-ins or electric cars would be the same as taking 60–90 percent of the cars off the roads of the United States and the world. In the United States this is equivalent to saving the potential oil in the Arctic National Wildlife Refuge every year.

All-electric cars would reduce $CO_2$ emissions by 900 million tons a year. We eliminate a large, significant chunk of $CO_2$ emissions just by switching to a new propulsion system for the nation's cars. In cities, hybrid plug-ins could reduce auto emissions to near zero, since most car trips are short and wouldn't require the car to use the internal-combustion back-up engine. The production of efficient, reliable hybrid plug-ins is now possible with the present state of technology. However, work needs to be done to further improve the energy density and charging systems of batteries.

## Electric Cars

The internal combustion engine is dead. Make no mistake about it. The internal combustion engine will be a relic, an antique, a quaint reminder of the past. It will go the way of the horse and buggy, the Stanley Steamer, and the tube tire. We are witnessing the "extinction" of a mechanical marvel, but many simply refuse to acknowledge its inevitable passing. I've loved it all my life, and many will lament its passing, but my grandchildren's children will remember it as a complicated mechanical contraption. It is dead because it no longer makes sense. Even its most efficient configurations are still dirty gas hogs. Small and smaller internal combustion engines will enjoy an extended life in hybrids and hybrid plug-ins, but they too will disappear as light, all-electric cars will take over the market. Of course, the electricity used must be produced from renewable sources, and not from fossil fuels.

> We eliminate a large, significant chunk of $CO_2$ emissions just by switching to a new propulsion system for the nation's cars.

### A Quick History of Electric Cars

The earliest cars were Battery-Electric Vehicles (BEVs). Built as early as the 1830s, they flourished for a while. Early in the twentieth century, BEVs out-sold gasoline-powered vehicles. However, after the invention of the radiator to keep an internal combustion engine cool and the introduction of an electric starter to replace hand cranking, the electric car could no longer compete. By the late 1930s, the electric car completely disappeared.

The invention of the transistor in 1947 brought new life to electric vehicles (EVs), but this new life was short. In 1961 production of EVs again stopped. This vintage of electric cars could reach speeds of more than 60 miles per hour and travel for an hour on a single charge, but they were too expensive to build.

Despite their "extinction," BEVs were always inherently *far more efficient* than the internal combustion engine (ICE). For example, a gasoline-powered ICE is approximately 20–25 percent efficient, and a diesel engine is approximately 30–40 percent efficient; however, an alternating-current induction electric motor is over 90 percent efficient. BEVs also *outperform* ICEs. The peak torque curve of BEVs starts close to zero revolutions per minute (rpm) and stays almost flat over the entire rpm range. Such performance is simply not available with even the most sophisticated ICEs. Further, BEVs produce neither $CO_2$ nor noxious emissions. When charged by electricity generated from nuclear or other renewable energy sources, BEVs create virtually *no pollution.*

The killer flaw in early BEVs was the battery. They cost a lot, provided very limited driving range, and took a long time to charge. All of these shortcomings are now disappearing. Many are adequate now. As of July 2006, between 60,000 and 75,000 battery-powered vehicles were in use in the United States.

## The Past Is the Future

Electric motors have been around a long time and are mechanically simple, inherently reliable, and extremely efficient—far more efficient than ICEs. The replacements for ICEs will be all-electric cars and hybrids with household plug-in capability to charge the batteries that go 250 or more miles between charges. We know how to build these vehicles, and some are now for sale on a limited basis. These developments must accelerate.

A gallon of gasoline used to generate electricity for an electric vehicle takes you at least 2 times farther than the same gallon of gasoline takes you in a car with an internal combustion engine. The electricity costs a few pennies per mile, yet gasoline costs approximately 12¢ per mile in the average car, both before taxes. Why bother arguing about whether global warming is severe or why it is happening? Why argue over fossil fuel's merits and problems? We can reduce transportation emissions now and ignore the arguments.

### The Near Future

The Discovery Channel broadcast an amazing program called *Future Car*. Because computational capabilities of computers double about every 2 years, automotive computers will soon be able to control … well … everything. Sensors will essentially eliminate accidents. Sensors will manage traffic and minimize congestion by spacing moving vehicles only inches apart. In fact, computers will operate cars better than humans could. Sounds great. Oh, but we must sustain ourselves in order to get there.

As improvements to hybrids and hybrid plug-ins are made, the internal combustion engines in hybrids will be used less. Of course, hybrids and hybrid plug-ins are a valuable stepping stone to all-electric vehicles, but they still consume precious oil and continue to pollute.

Plug-in hybrids have zero emissions most of the time. However, this claim holds true only if nuclear energy or other renewable energy sources generate the electricity. With plug-in hybrids, drivers can run most day trips on battery power only and reserve the gas engine for longer trips.

### The Future Is Now: Tesla and Phoenix

Most components required for battery-operated electric vehicles are fairly mature and cost competitive. The batteries, and to some extent the chargers, will determine how fast the world adopts BEVs. The rate of adoption will depend on the cost, availability, and ability to recycle batteries. Lithium-ion, lithium-ion polymer, and zinc-air batteries have demonstrated energy densities high enough to deliver reasonable driving ranges and recharge times (as low as 10 minutes using specially designed charging systems) comparable to filling the gasoline tank of a regular car.

Tesla Motors, a small company, is one of the most advanced electric-car manufacturers in the world. A Tesla Roadster (see Figure 14.3) can go over 200 miles between charges. Each charge takes 3.5 hours from ordinary house current, but it is always charged and ready to go if kept plugged into your house current when the car is not in use. Tesla cars use recyclable lithium-ion batteries to store the charge, the same batteries used in laptops, cell phones, and other consumer electronic devices. A set of batteries is predicted to last over 100,000 miles. According to Tesla, if one uses off-peak current at 5¢ per kilowatt hour, then the fuel cost would be approximately *1¢ per mile*. This, of course, does not include taxes. Compare this to the 12¢ per mile pre-tax fuel cost of a conventional car. The average gasoline tax in the United States is about 47¢ per gallon, or about 2.5¢ per mile.

The Tesla motor is over 90 percent efficient and weighs about 200 pounds. Tesla's all-electric car is efficient to operate, and it is a screamer—faster than just about any thing else on the road—0–60 miles per hour in about 4 seconds. Tesla's motto is "Burn rubber not gasoline."

All in all, Tesla's superb car is priced well below other luxury sports cars such as the Ferrari Enzo or the Lamborghini Diablo, and will equal or surpass their performance. A Tesla sedan is coming, too. Tesla planned to produce 10 roadsters per month in 2007 and plans 100 cars per month in 2008. I'm not surprised Teslas are sold out until late 2008. Besides being clean and efficient, they are beautiful cars built by the Lotus Car Company in England.

**Figure 14.3. Tesla Roadster**

What about service? Probably expensive, right? Wrong. Service costs are far less than gasoline-powered cars. A Tesla car has no oil or oil filters to change, no air filters to replace, and no smog equipment because the cars emit no exhaust fumes to filter or reduce. The only service required will be to inspect and maintain the brakes, suspension, steering, and tires. Many service expenses essentially go away.

The car costs $90,000–$100,000 depending on accessories. As with all new production, costs will likely go down as more units are built. Consider also that over 100,000 miles of driving you will save over $10,000 in fuel costs and some in service charges. And you will never have to pay a gas-guzzler tax.

According to Tesla, if one uses off-peak current at 5¢ per kilowatt hour, then the fuel cost would be approximately 1¢ per mile.

Another innovative small company, Phoenix Motor Cars, produces sport-utility trucks (SUTs) that go 135 miles between charges (see Figure 14.4), and they can be recharged in only *10 minutes* using a special charger costing about $30,000. (Phoenix plans to bring a companion sport-utility vehicle to market in 2008.) This handsome truck can be purchased for the very competitive price of approximately $45,000. How much did you pay for your last gas-guzzling SUV or pickup truck? The fuel cost to drive the Phoenix SUT is 10 percent of the cost to drive a typical SUV. Also, Phoenix uses a nano-enhanced lithium-ion battery. This battery permits rapid charging and is said to be safer than other lithium-ion batteries. In addition, the recyclable battery will have a working life beyond the life of the vehicle. Phoenix plans to build over 20,000 vehicles in the next 3 years.

Thank you, Tesla and Phoenix. The country owes you a debt of gratitude for showing us the way to pollution-free transportation. You've shown us the way to the future. Batteries will get better, the range between charges will be extended, and charge times will be reduced—you can bet on it.

Hybrid plug-ins are available today only as after-market add-ons. They should be available soon directly from some of the big car companies. How about it GM and Ford? GM announced that the Chevy Volt will be ready by 2010 or so. GM is hyping the Volt now to prove the company is in tune with fuel realities. Performance tests have been excellent. If GM cannot deliver the Volt sooner, GM risks remaining a follower.

> A Tesla car has no oil or oil filters to change, no air filters to replace, and no smog equipment because the cars emit no exhaust fumes to filter or reduce.

**Figure 14.4. Phoenix truck**

*BEVs in Other Countries*

- **France**. Citroën Berlingo Electrique built several thousand BEV delivery vans, mostly for fleet use in municipalities and by Électricité de France. Production and use of these vans is almost completely clean since most of the electric energy used to produce and operate them was/is nuclear.

- **Norway**. BEVs are tax exempt and can use the bus lanes.

- **Switzerland**. BEVs are popular with private users, and there is a national network of public charging locations called Park & Charge. There are also some charging stations in Germany and Austria. The Swiss always seem to make sense.

▶ **United Kingdom**. Electric vehicles in London are exempt from certain taxes. Most cities use low-speed electric milk trucks to deliver fresh milk to homes.

▶ **Italy**. BEVs are exempt from taxes and enjoy substantial reductions in insurance fees. In many cities the trash collection is performed by BEV trucks.

▶ **India and China**. These countries are starting to build BEVs. India has built over 1000 since 2001 and exports some.

## Coming Down the Road

Some major and minor automobile manufacturers are hinting at coming breakthroughs in the availability of plug-in hybrids and electric vehicles suitable for everyday use in general traffic. Here are some of the hints:

▶ California Cars Initiative, E drive Systems, HybridsPlus, and Hymotion—These small companies plan to produce plug-in hybrid electric vehicles in the near future. By adding more battery capacity, these firms expect to attain 250 mpg between plug ins.

▶ Mitsubishi—The Mitsubishi Corporation is committed to creating a flexible-fuel vehicle that it can produce in alternative forms: as a BEV, a hybrid, or as a fuel-cell vehicle. According to a report on the MSNBC cable television network, Mitsubishi plans to unveil its electric car in 2010. According to *AutoWeek* and MSNBC, Mitsubishi's Colt is expected to have a range of about 100 miles between plug-ins using lithium-ion batteries and electric motors attached directly to the wheel. The target price is about $19,000.

- ▶ Société de Véhicules Electriques (SVE)—This French firm plans in 2008 to mass produce only for the French market an all-electric car based on the Renault Rangoo.

- ▶ Toyota—Ken Thomas of the Associated Press reports in the *Chicago Sun-Times* (July 19, 2006), that Toyota officials hint that the company's next generation hybrid will have lithium-ion batteries with a 9-mile battery-only range that results in over 100 mpg in certain conditions. I hope Toyota is measuring the miles per gallon under normal driving conditions and not under the cooked books used by the Environmental Protection Agency's industry-friendly mpg ratings. For example, the EPA inflated the original Prius's ratings by almost 30 percent.

- ▶ Subaru—In response to high and rising fuel prices, Subaru may accelerate development of its R1e electric car. Subaru's advanced in-house battery technology and Toyota's hybrid Synergy Driver technology will be shared by both companies so both will benefit. A win-win for both automakers.

Consider this exciting news about Subaru's R1e, as reported online by *Megawatt Motorworks*.

> Subaru Canada, Inc. (SCI) today unveiled the R1e proto-type electric vehicle at the Canadian International Auto Show in Toronto. The R1e, which is based on the R1 mini-car currently available in Japan, was created using new manganese lithium-ion (Li-ion) battery technology that means the vehicle could recharge to 90 per cent of its capacity in just five minutes with the use of an exclusive charger.
>
> This means that recharging the R1e could be done almost as fast as completely refueling a gasoline pow-ered car, which typically takes about three minutes. It's

certainly a significant improvement over other electric vehicles, many of which can take eight hours to recharge. The R1e is designed to go about 120 kilometres [80 miles] on a full charge. The new Li-ion batteries are also extremely small and lightweight—the R1e's battery pack is about the size of a VCR—and designed to last for 10 years or 150,000 kilometres [100,000 miles]...

These high-technology vehicles are part of Subaru's broad approach to environmental responsibility that takes into account vehicle emissions and fuel efficiency, as well as reduced environmental impact from all facets of automobile development, production and marketing.

In addition, FHI [Fuji Heavy Industries] is currently conducting performance tests on prototype cells of a new Li-ion capacitor with enhanced power density (instantaneous force) and high energy density (cruising distance) for a next-generation vehicle that meets goals for sustainability without sacrificing performance. The successful commercialization of Li-ion capacitors for compact cars would open up many other business opportunities, including helping to meet the increased demand for alternate fuel buses, trucks and other passenger vehicles. This new capacitor also has potential as an alternative to conventional lead-acid batteries in the future.

Other companies are working on plug-in hybrids and BEVs. Ford, Honda, Lexus, Pininfarina, Volvo, Smart, and Suzuki are some of them. *The race is on, and that is good for all of us.* I'm sure this race will realign the automobile industry. The biggest winners: the early adapters, the environment, our children, and grandchildren.

### Assault on Batteries

In the future, batteries will perform in larger and more important applications. They will propel our cars, buses, and trains. Batteries will render wind and solar energy much more efficient by providing high current density storage of electricity, when the grid does not need the energy being generated when the wind is blowing or the sun is shining. Battery technology enables and improves many energy-related technologies. The future belongs to batteries.

We must mount a man-to-the-moon research blitzkrieg to develop better batteries that are long-lived, recyclable, economical to manufacture, and provide greater current-carrying capability. Much research is being done now—but not nearly enough. Many researchers around the globe hope to apply nanotechnology—the engineering of functional systems at a molecular scale—to increase battery cycles and current density 20 times higher than the batteries used today. Such developments could conceivably permit a car to go 500 plus miles between plug-ins. I can hardly wait. Companies will have to get into the game or wave goodbye from the sidelines.

> The future belongs to batteries.

Battery-electric vehicles have used many types of batteries, including lead-acid, nickel-cadmium, nickel/metal hydride, lithium-ion, lithium-ion polymer, and, less frequently, zinc-air and molten salt. Batteries are getting better and better, but progress has been slow and should be accelerated. In the interest of time, the government should devote at least $1 billion to research directed to developing new and improved batteries and to investigating how to apply nano-science and other cutting-edge technologies to battery development. As with most technologies, the engineering challenge is to create the proper balance of characteristics: range vs. performance, range vs. charge time, and battery capacity vs. weight, for example. Attention to small advantages becomes important.

> Companies will have to get into the game or wave goodbye from the sidelines.

As an example, using braking energy to recharge the battery can extend the range of a battery-electric vehicle from 10–50 percent depending on driving conditions.

### Lithium-ion Batteries

Many lithium-ion batteries—over 6000 of them—are innovatively connected to power the Tesla car discussed above. Tesla engineers chose these batteries for several compelling reasons: their availability, long life, high current density, charging ease, light weight, ability to recycle, and the ability to configure them in a car for optimum weight distribution.

Further, because lithium-ion batteries are important to the electronics industry, a great deal of research has been and is being done to improve them. In fact, in recent years their energy-storage capacity has improved by an impressive 8 percent annually. If this rate of improvement continues, their capacity will double in fewer than 10 years, thereby permitting the Tesla car to travel twice as far on a charge. This could mean 400 miles per charge.

### Nanotechnology Batteries

Phoenix Motor Cars use NanoSafe™ batteries from Altairnano, Inc., another innovative company of note and a leader in advanced nano-materials and alternative energy solutions. Altairnano officials believe their technology represents a giant leap forward in the design of batteries for electrical vehicles. Altairnano has developed novel electrode nano-materials, and its rechargeable, nano-titanate NanoSafe™ battery system provides fundamental advantages over current lithium-ion battery designs.

Altairnano's goal was to develop electrode materials to enable a battery to be charged in minutes, deliver high power, and be long lived. Fundamental research on the electro-chemistry of battery materials led to the conclusion that nanotechnology could provide dramatic new material properties that could enhance these parameters. Altairnano's engineers, already possessing substantial knowledge of nano-titanate materials, postulate that replacing graphite in conventional lithium-ion batteries with nano-titanate materials would result in batteries that solve issues of charge time, lifecycle, power, and safety.

### Batteries Affect Range

The range of an electric vehicle depends on the energy-carrying capacity of the battery. However, battery choice must be subject to some practical limitations such as weight and volume. As a result, the current density (expressed in energy per pound or energy per cubic inch) becomes all-important. The lead-acid batteries used in some hybrid plug-ins have a range of 20–80 miles, yet lithium-ion batteries achieve ranges of 250 miles or more.

### Charging Ahead

Batteries must be charged (or recharged) from time to time. Since the time and ease of charging will make a big difference in the deployment of BEVs, charging is a fertile area for innovation and improvement. Obviously, a vehicle's range between charges becomes much less important if the battery can be charged quickly. However, most current batteries have a fixed, specific charge rate and cannot be charged faster. Most people would prefer to recharge their batteries at home at night when electricity rates are usually lower. Charging does not need attention and plugging in takes only seconds. As all-electric vehicles become more popular, work places, hotels, schools, restaurants, and shopping centers will provide facilities with charging equipment.

In 2006 Subaru demonstrated its R1e car can recharge to 90 percent capacity in 5 minutes using an exclusive charger. In 2007, Altairnano's Nano-Safe battery could charge to 80 percent capacity in about 1 minute and is fully rechargeable in a few minutes. This industry is well on its way.

## Retrofit Kits

Converter kits on the market can change some hybrids into plug-in hybrids, and other kits completely replace an internal combustion engine with batteries to convert the car to an all-electric vehicle. A popular battery for these retrofit kits is the deep-cycle, lead-acid battery. Retrofitted cars with these batteries could have a range of up to 80 miles. Still, a range of even 40 miles would satisfy most drivers for most short, around-town trips.

I strongly encourage such innovations and experiments. We all know that this kind of thinking can produce amazing results. These early, in-their-own-garage innovators believe all things are possible. And they often prove it. That's American ingenuity at its best.

## FOOT-DRAGGERS, RESISTERS, and OPPONENTS

An all-electric car transportation fleet will deliver a big blow to those who hang on to and earn their livelihoods from the internal combustion engine. The advantages of no gasoline, no internal combustion motors, no complicated transmissions, no cooling systems, no mufflers, no catalytic converters, and no exhaust systems are overwhelming. Will the owners of gasoline stations, muffler shops, and repair garages embrace the changes? Will the manufacturers and factory workers producing conventional batteries, oil filters, air filters, radiator hoses, drive trains, transmission fluid, and the diverse other components and accessories of internal combustion engines eagerly embrace the changes? Of course not. Expect reluctance and resistance. Technical changes will come, and they are as inevitable as the opposition they will spark.

Be very skeptical of people with vested interests criticizing new technologies that threaten their interests. It is natural for people to protect their positions and defend their turf. However, these self-interested criticisms, protections, and defenses are often not in the broad interests of the general public. They can cause delay, sow confusion, and inhibit progress. Delay can lead to oil depletion, resource wars, and an earth that becomes less and less habitable as it becomes more and more toxic.

Many companies and their employees will suffer unless they embrace the new technologies and find opportunities there. They can't win or survive by hanging on to old technologies any more than the horse and buggy people could endure the onslaught of the automobile. Come to think about it, how far could a horse travel before it had to be recharged?

Consider for a moment the global market for batteries. No oil company or automotive company yet controls the lithium battery market. Lithium-ion batteries were developed by East Asian and Canadian firms for use in portable computer equipment. That means the patents and production are currently beyond the legal and financial reach and control of U.S. automakers and oil companies, rendering them powerless to stop this technology.

## POLITICAL ISSUES

Never underestimate what large political contributions to political parties and candidates can achieve. Entrenched industrial interests will seek regulations that suppress potential business competition. We understand the self-interest at work. But the public has interests, too. Entrenched industrial interests will work through political lobbyists and advocate their (self-interested) positions under the guise of some broader virtue, such as protection of consumers, of electricity grids, and of jobs in the automotive industry. Such claims remain a threat to competitive markets and to a sensible energy policy in the United States.

I recommended that you see the movie *Who Killed the Electric Car?* The subject was General Motors and its all-electric car. The reality, however, is that nobody can kill the battery-electric vehicle, because BEVs just make too much sense. It will be fun to watch progressive companies roll out their own unique versions of future automobiles. Still, the public must help where it can.

We must take some risks, which is a problem in a country where the legal profession has taught us that a failed risk is likely a lawsuit.

This must change since a riskless society eventually becomes a second-rate, sickly, morbid society. Pioneers, adventurers, and entrepreneurs embrace risk. We cannot execute a rational energy plan without some risk, and to do that we need freedom from senseless legal road blocks.

Why do I make these political points in a book about energy? Because energy is political. Because energy policy is deeply political. Because *every single scientist* I asked about the lack of sane laws, policies, and technologies to deal with our huge energy problem answered me with one word—"POLITICS." Politics as usual. The politics of special interests. Corporate politics. The politics of privilege, power, party advantage, and big bucks prevailing over common sense and common people. Some were keenly aware that if they did not support the current political "scientific" agenda, then they would lose their funding. Big money can bend even science to do its bidding.

> Technical changes will come, and they are as inevitable as the opposition they will spark.

Why do I spend so much time on automotive fuel performance and the politics of American car companies? I do this to caution against making the same mistakes in the future. Further, I believe a vital American auto industry will help us achieve energy independence sooner. GM, Ford, and Chrysler command a large part of the American auto market, and can be a large part of the solution.

## BOTTOM LINE

An exciting, clean, healthy, and profitable future awaits us when we mobilize and "electric-motorize" the future. Benefits: lower transportation costs, less imported oil, improved balance of payments, a stronger dollar, fewer resource wars, less pollution, and more will result from the new realignment of the auto industry and energy sources.

What must we do now?

◖◗ **Rapidly move to hybrid plug-ins and all-electric cars.** This is essential to our march to energy independence by 2040. Since this is a "critical path" to our goal, we must spare no cost or effort to succeed.

◖◗ We must strive to have **90 percent of our automobile fleet be plug-in hybrids or all-electric vehicles by 2040.**

◖◗ **Speed up research and development of batteries and battery-charging systems** suitable for use in electric vehicles.

◖◗ **Phase-out gasoline-powered vehicles.** Establish a legal maximum on the number of such vehicles that can be sold each year, and aggressively reduce that number each succeeding year. Little hardship was caused when we couldn't buy new cars during World War II. The public can accelerate the transition by simply not buying new cars until hybrid plug-ins and all-electric cars become available.

◖◗ **Add a surcharge of 50¢ per gallon of gasoline.** The surcharge must be fully dedicated to developing and deploying clean, renewable energy and transportation systems. More on this in Chapter Sixteen.

◖◗ **Charge customers a surcharge** of 5 percent of the vehicle's sticker price for cars getting less then 30 mpg, and levy a 10 percent surcharge for vehicles getting less than 20 mpg. In 5 years the surcharge should be increased: 5 percent surcharge for cars under 40 mpg, 10 percent for cars under 30 mpg, and a 20 percent surcharge for cars under 20 mpg. Again, see Chapter Sixteen for details.

◖◗ **Ration gasoline.** Rationing may also help to stabilize prices.

◖◗ **Deploy plug-in stations at convenient locations.** Service stations could sell electricity instead of gasoline for fast charges. Retail companies might find it good business to install charging stations. Hilton Hotels has installed charging stations at three of its hotels. Companies should be mandated to provide charging outlets for their employees.

▷ **Increase the penalties manufacturers pay for non-compliance with efficiency standards.**

▷ **Support GM, Ford, and Chrysler** as they get back in the game. Besides, more competition begets more innovation, and innovation is exactly what the world needs now, more than ever.

## Chapter 15

# BRIDGING THE GAP

## TO ENERGY INDEPENDENCE BY 2040

A tremendous gulf separates our present situation from the sustainable-energy future I envision. The gulf is deep and wide. A future of clean, affordable, renewable, environmentally sound energy must be attained by 2040. That gives us a bit longer than a single human generation—3 national censuses, 7 summer Olympics, the opportunity for a newborn to grow, start a family, settle into a career, and vote in 3 presidential elections. I recommend an aggressive 30-year transitional plan to get us to that future. In short, the United States needs a "bridge" to span the gulf that separates our present energy woes and economic dependence from our future energy independence and economic prosperity.

Every individual, government, corporation, organization, and school should adopt this slogan: "Energy Independence by 2040." A longer transition is simply too dangerous, as we shall see.

The transformation to clean electrical energy and all-electric

> Every individual, government, corporation, organization, and school should adopt this slogan: "Energy Independence by 2040."

vehicles will take decades. If the U.S. public procrastinates, then we'll likely need 50 years—but 50 years is too long. What kind of a bridge is necessary for a 30-year transition?

**30-Year Transition:**

- **Oil.** The United States needs 175 billion barrels of oil. With only 20 billion barrels in reserve, the United States must find or buy 155 billion more barrels. At a cost of $100 per barrel, the total will be $15.5 trillion dollars. However, over the 30-year transition, the price of oil will likely increase to $200 per barrel or more. Over the same 30-year period, the world will need approximately 700 billion barrels, and, as a result, fierce competition will erupt for the world's dwindling supply of oil.

- **Natural Gas.** In 30 years the U.S. reserves of natural gas will be dangerously low. At current consumption rates, the reserves will last only about 65 years until bone dry.

- **Coal.** Coal reserves are okay.

- **Nuclear.** In the next 30 years U.S. utilities must build a minimum of 400 nuclear plants—ideally, 800 plants. Progress will be slow initially, but it must accelerate over the 30 years. Rapid implementation would, of course, relieve pressure on natural gas resources and help conserve coal as a feed stock for plastics, chemicals, and more.

- **Transportation.** In the next 30 years the United States and the world must move rapidly to biofuels, hybrid plug-ins, and all-electric vehicles. This is a crucial point. The more rapidly the world moves to these fuels and vehicles, the longer oil reserves will last.

Since it is unlikely that the United States will be able to import the quantity of oil it needs in coming years, it is essential—indeed imperative—that the U.S. public vigorously support the extraction of oil from vast U.S. oil shale deposits. Also, by 2040 cellulosic ethanol and algae biodiesel must supply most of the liquid fuel required for hybrid plug-ins and other U.S. transportation needs. *These are the key features of the proposed bridge: Move rapidly to hybrid and all-electric vehicles to reduce the need for oil, develop needed oil from domestic oil shale resources, develop biofuel alternatives to gasoline, and generate all new electricity needs from renewable energy sources.* We must become the masters of our own destiny, not dependent upon essential resources controlled by others.

For a dire comparison, consider a potential 50-year transition.

## 50-Year Transition:

> We must become the masters of our own destiny, not dependent upon essential resources controlled by others.

- **Oil.** The United States needs 420 billion barrels. Over the same 50-year period, the world needs at least 1.6 trillion barrels, but probably needs closer to 2 trillion barrels. But here's the problem: The world has only 1.1 trillion barrels of conventional oil left. The extra oil must come from unconventional sources. But I have little confidence that the world can recover enough unconventional oil in time.

- **Natural Gas.** The United States and the world will face serious shortages as world reserves approach total depletion.

- **Coal.** Coal reserves are okay, unless coal is used to make gasoline as oil supplies dwindle.

You can readily see that a 50-year transition would be extremely difficult. The world can and must make this transition within about 30 years—that is, by 2040. All nations of the world must be encouraged to adopt the same 30-year transition plan.

## FALSE BRIDGES

We must avoid "false bridges"—options that look or sound good only when promoted by special interest groups, or options that will take too much time and effort yet will still fall short. Such exercises could cause delay deploying a real solution, delays the world can ill afford. During this 30-year plan *the United States must NOT*:

- ◗ Drill in the Arctic National Wildlife Refuge (ANWR), in the National Petroleum Reserve of Alaska (NPRA), or in the Beaufort Sea because of the potential environmental harm, and because there is not enough oil there.

- ◗ Drill offshore because it is environmentally risky, and the total reserves are likely to be woefully short of the amount needed.

- ◗ Invest in the so-called hydrogen economy.

For several reasons—the extent and significance of the area to be disturbed, the timetable, and the costs—extracting and refining oil shale is a much better alternative to the options listed above. I believe extracting and refining oil shale is the *only* workable solution in constructing the necessary bridge.

### False Bridge 1: Drilling in Alaska

An unnecessary environmental tragedy could unfold if the U.S. federal government permits oil companies to drill for oil in ANWR, in NPRA, or in the Beaufort Sea—there is not enough oil in these places to be much of a bridge. Drilling would be a waste of time, effort, and money better directed to producing oil from oil shale.

According to Joel Bourne in *National Geographic*, "In the petroleum-rich wilderness Alaskans simply call it 'the slope,' where big money, power politics, and hype run as thick as mosquitoes, oil companies have paid $120 million to lease nearly 2 million acres." What a bargain, since $120 million would not come close to paying for environmental damage that would result with wholesale drilling on this land, most of which belongs to the citizens of Alaska and to you and me. The *National Geographic* article further declares:

> Most of our [U.S.] holdings are split between the scenic ANWR and a 23-million-acre chunk of western arctic known as the National Petroleum Reserve, Alaska, or NPRA. The NPRA contains the largest piece of unprotected wilderness in the nation, along with a half million caribou, hundreds of grizzlies, wolves, and in summer more waterfowl, raptors, and shorebirds than anyone can count.

Biologists have argued for decades that areas of NPRA are more critical to wildlife than the ANWR. While hard data on human-induced global warming has been elusive, the damage that could be caused by wholesale drilling on the North Slope is accurately predictable based on experience.

*National Geographic* reports that "while federal and state biologists have been warned to hold their tongues while the battle over drilling in the refuge rages in Congress, the Bush Administration leased vast tracts of the petroleum reserve [NPRA] and offshore waters to the highest bidder."

I'm getting steaming mad as I write this. You, too? Government officials are muzzling the very people who know the most about the situation and selling leases to their friends who provide political campaign funding. Since when do politicians get to muzzle scientists or anybody else? Citizens simply cannot trust elected officials in Washington to consistently serve the public interest as long as the existing political system permits well-funded lobbyists and corporate warriors to buy off politicians and frustrate the democratic process. Money corrupts Washington, so corrodes and befouls our democracy, thereby condemning our nation to politics by the highest bidder and the deepest pockets. Is "follow the money" the simplest description of national politics? Has George Washington on the dollar bill become more important and influential than George Washington the man and his principles? Perhaps America has adopted a new set of heroic icons: the athlete in the movie *Jerry Maguire* who shouts "Show me the money." and the character Gordon Gecko, who ominously intones "Greed is good" in the film *Wall Street*. Have mountains of cash replaced Mount Rushmore as symbols of the national spirit?

OK, I've declared that not enough oil sits in ANWR and NPRA, and I've condemned the influence of corporate money in national politics and a political process that too often rewards the rich. Let's look at details.

### Environmental Damage

Some politicians, even some from Alaska, claim that the North Slope is a bleak wasteland good for nothing except the oil beneath the surface. However, ANWR, NPRA, and the Beaufort Sea are sensitive and important habitats, breeding grounds, and summer sanctuaries for many species of birds, marine life, and land animals. In addition, the area is also a vast feeding ground for marine life. Phytoplankton and marine algae—the bottom of the food chain—abound in ANWR, NPRA, and the Beaufort Sea like no other places on earth. An offshore oil spill would devastate these ecosystems. Should the rest of the world have the opportunity to voice views about this risk? Would anyone bet against an offshore spill, no matter what security systems and safety precautions are in place? A spill would upset not only the arctic ecosystem, but also ecosystems world-wide, since the genesis of much life on this planet begins in the arctic. Let's face it—oil spills are frequent, about 100 spills each year. In fall 2007 a large spill occurred in San Francisco Bay, where safety precautions are well-known. A different spill by British Petroleum totaled 200,000 gallons. BP earns almost $20 billion per year, yet can't seem to properly maintain its pipelines. Out of sight, out of mind? Clean up in remote areas can be practically non-existent. Federal agencies are still trying to clean up the contaminated soils at Umiat, Alaska, nearly 60

Oil spills are frequent, about 100 spills each year.

years after the U.S. Navy drilled there. Of course, in such remote areas, who's to notice?

Experts predict that recovering oil from ANWR and NPRA would cause irreversible damage to the animals and plant life. Recovering oil from oil shale also damages the environment, but the damage is reversible, and we know how to restore disturbed areas.

### Not Enough Oil in ANWR and NPRA to Make a Difference in an Energy Crisis

Estimates vary on the combined amount of oil in ANWR and NPRA. No matter how you count or what estimates you work with, the amount could provide only 5–15 percent of the "bridge" required for the 30-year transition plan. While it would provide great returns to the oil companies, any statistician or gambler will quickly tell you that the odds are stacked against the Alaska option providing much of a bridge. In a sane world it is no option.

### Mission Creep

If drillers encounter more than a bucket of oil in ANWR or NPRA, then they will apply tremendous pressure to drill in the Beaufort Sea to recover offshore oil. *Again, oil companies have already bought leases.*

### The Politics of Profit

After the Clinton Administration engaged in a lot of political posturing and policy pussy-footing, critics blasted the administration for treating Alaska as a "cookie jar" for oil companies to reach in to. The administration of President George W. Bush claimed oil exploration and drilling would have a lesser impact on wildlife than previously predicted. This claim—and the accompanying Environmental Impact Study (EIS)—bars oil exploration on only 6 percent of the Alaskan coastal plain. Has anyone seen this EIS? Some critical

> If drillers encounter more than a bucket of oil in ANWR or NPRA, then they will apply tremendous pressure to drill in the Beaufort Sea.

wags describe Bush's energy policy as "no oil man left behind."

## False Bridge 2: Drilling Offshore

Recovering oil offshore is a better option than drilling in Alaska, but still an insufficient and unnecessarily risky choice. Why risk environmental calamity by drilling offshore into deeper and deeper waters when safer and cheaper options exist? Offshore drilling in the United States makes little sense because it costs approximately the same as recovering oil from oil shale and is very much riskier. This startling claim is from a 2004 report entitled *Strategic Significance of America's Oil Shale Resource, volume II*, a report prepared by the Office of Deputy Assistant Secretary for Petroleum Reserves of the U.S. Department of Energy. Yes, the potential profits are a strong incentive, but there are safer profits to be made in oil shale. Offshore drilling only makes sense for countries that lack other resources and options. Oil deposits are uncertain, but even the maximum estimates are insufficient for the proposed 30-year bridge.

## False Bridge 3: Hydrogen

Forget the so-called hydrogen economy, as detailed in Chapter Thirteen. For now, simply put hydrogen out of your mind. One must produce hydrogen just as the world currently produces electricity. If we burn polluting fossil fuels to produce clean-energy hydrogen by separating water into oxygen and hydrogen, then there is no environmental benefit. Finally, producing hydrogen requires more energy than is contained in the hydrogen produced.

## Oil Shale Wins

By process of elimination and by virtue of its merits, extracting oil from oil shale triumphs over drilling in environmentally sensitive

ANWR and NPRA, where not enough oil sits anyway. Oil shale also trumps offshore drilling and the fantastic allure of a potential hydrogen economy. Ignore these temptations.

Only oil shale offers a reasonable, reliable, available solution. Oil shale reserves are very large, reasonably economical to recover, and would surely provide the United States an adequate bridge to a future of clean, renewable energy. No matter what we do in Alaska, we still need oil from oil shale for a stable energy bridge into the future. So why not turn to oil shale right away? At a production cost of even $60–80 per barrel, oil shale is economical to recover. However, according to *Strategic Significance of America's Oil Shale Resource*, the cost to recover oil from oil shale will be closer to $20 per barrel—but more likely to be $25–50 per barrel. Besides, oil shale is not located in sensitive wildlife areas.

Before turning in detail to the specific merits of oil shale, let's first look at Canada's production of oil from oil sands. Canada's efforts show the world the way.

## CANADIAN OIL SANDS

Let's review: We can and must immediately reduce our use of coal to generate electricity. We must replace coal and other fossil fuels with nuclear power and other renewable energy sources to generate electricity. However, we need time to solve our transportation needs and slash the huge volumes of petroleum our vehicles guzzle. Therefore, we need to find oil we can use for the next 30 years—or longer, if need be. Thus, the United States and the world at large will need to recover oil from unconventional sources. We can partly fill our transportation needs with ethanol and biodiesel supplemented with conventionally produced oil—but that falls well short of providing enough of a bridge. That brings us to oil sands and oil shale. Industry experts consider oil shale, oil sands, and heavy crude oil as "unconventional" sources.

Tar sands, which our Canadian neighbors prefer to call "oil sands," are a cousin to oil shale. I learned much about oil sands during interviews with incredibly cooperative Canadian government officials and oil industry experts. I am particularly grateful to Patti Lewis of Suncor, who was especially knowledgeable and helpful. (For comparisons of oil sands and oil shale, see Figure 15.1. You will note that U.S. oil shale compares quite favorably with Canadian oil sands in most evaluation criteria.)

**Figure 15.1. Comparison of Principal Factors Influencing the Economics of Unconventional Crude Oil Production**

| Characteristics | Canadian Oil Sand | U.S. Oil Shale |
|---|---|---|
| Resources | More than 1 trillion barrels | More than 1 trillion barrels |
| Grade (Richness) | 25 gallon bitumen/ton | 5 gallon kerogen oil/ton |
| Hydrogen Content (bitumen/raw oil) | 10.5% by Weight | 11.8% by Weight |
| Nitrogen & Sulfur Requiring Removal | 6.2% by Weight (mostly Sulfur) | 4.0% by Weight (mostly Nitrogen) |
| Loss of Liquids to Coke and Gas | 40 pounds/ton-ore | 11.6 pounds/ton-ore |
| Net Yield of Oil | 0.53 barrels/ton processed | 0.60 barrels/ton processed |
| Quality of Oil | 34° API | 38° API |

Source: U. S. Department of Energy, Office of Naval Petroleum and Oil Shale Reserves, *Strategic Significance of America's Oil Shale Resource*, volume II.

Processors can mine oil sands and truck the "ore" to a plant for processing, or they can heat the sands in the ground and extract the oil with a conventional oil rig. Although engineers working with oil sands and oil shale have learned much from each other, oil shale appears to present greater processing and financial challenges than oil sands.

Canada presently produces approximately 1 million barrels of oil per day from oil sands and will spend $65 billion to increase capacity by 2030 to almost 5 million barrels of oil per day, which equals 25 percent of current U.S. daily oil use. Total production cost per barrel of oil from oil sands, including the capital invested, is $20–25 per barrel.

With present technology Canada can potentially and economically recover over 175 billion barrels of oil, which is about 10 percent of its massive oil sand resource of 1.7 trillion barrels. New, more economical technologies are almost certain to come and will permit even greater recovery at lower costs. Indeed, the costs of producing oil from oil sands will surely fall. Canadian engineers have developed a new, state-of-the-art technology for surface processing (retorting) of oil sands. The process, called the Alberta Taciuk Process (ATP), increases the yield of oil and combustible gas, improves thermal efficiency, and reduces water requirements.

Anyone experienced with new products and processes would quickly agree that costs always fall as engineers and scientists discover or innovate new ways to reduce costs. The oil sands industry in Canada is a fine example: It has reduced the cost of a barrel of oil from oil sands by about $5 per barrel in the last 10 years, and Canadian processors are confident of significant cost reductions in the future.

The Canadian government provided a key incentive to develop the oil sands industry. The government forgives oil sand royalties (the government owns most oil sands reserves) until project payback is achieved. No other incentive exists. However, those I interviewed said the Canadian government has done everything possible to clear obstacles so the oil sands industry could prosper.

I believe the Canadian government sees initiatives to solve the world's energy problems as a hugely attractive business opportunity, not as an occasion to make it difficult to develop alternative oil sources or to be beholden to Big Oil interests and unfriendly suppliers. Quite clearly, the Canadian government paves the way for progress and is careful not to obstruct progress or permit others to erect hurdles. This position comes through loudly and clearly when one reads the professional journal *Industries for Renewable Energy and the Environment*. One can't compare Canada's oil sands industry to the U.S. oil shale industry—because no U.S. oil shale industry exists. No comparison.

## U.S. OIL SHALE

While the United States is doing very little to develop U.S. oil shale resources, in spite of having the largest deposits in the world, others are aggressively exploiting oil from oil sands and oil shale. Wake up, America. The resources for a transitional bridge are right here in the USA. We only have to dig. Also, if we are to help other nations become energy independent, we should plan on providing them oil from oil shale at a reasonable cost because all nations will have the same problem of not having a sufficient bridge to energy independence. U.S. deposits contain the most extensive and economically recoverable oil from oil shale on earth.

Oil shale is a hydrocarbon contained in porous oil-bearing rocks. The hydrocarbon, called *kerogen*, can be converted to oil through a chemical process. Oil shale can also be burned directly as a low-grade fuel. Indeed, because the rock can actually ignite, it is sometimes called "the rock that burns."

Approximately 2 trillion barrels of recoverable oil sit in U.S. oil shale deposits, a total roughly equivalent to all conventional oil ever discovered. A significant portion of oil from oil shale can be recovered for $25–40 per barrel, depending upon on-site circumstances. We can produce oil from oil shale right now; we know enough about these resources and the relevant technologies to begin producing as soon as a plant is built. Why the United States has not developed this resource in the sure knowledge of conventional oil's rapid depletion is a testament to greed, misleading oil industry propaganda, political campaign financing, and political incompetence, sprinkled with a good dose of confusion and ignorance.

> Approximately 2 trillion barrels of recoverable oil sit in U.S. oil shale deposits, a total roughly equivalent to all conventional oil ever discovered.

Oil shale is located in places in the United States less ecologically sensitive than most locales, and the sites can be restored after the oil is removed. The largest and most economically attractive oil shale deposits are in the western United States. Found in semi-arid regions of Colorado, Utah, and Wyoming, the deposits are known collectively as the Green River Formation, located in the river basin of the Green River. Nearby towns include Grand Junction, Meeker, Rangely, Rifle, Rock Springs, and Vernal (see Figure 15.2).

## Figure 15.2. Map of Green River Formation Oil Shale and Its Main Basins

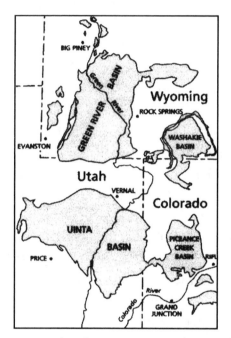

Here is what Dan Denning said about the area.

You won't think much of Rio Blanco County if you ever drive through it. In fact, unless you take a right turn off Interstate-70 West at Rifle, head north on Railroad Avenue and then west on Government road to Colorado state highway 13, odds are you'll never even step foot in Rio Blanco County.

But even if you keep heading west toward Grand Junction, through the town of Parachute and the shuttered oil shale refineries from the 1970s, you'll see the Book Cliffs geologic formation on your right. For miles and miles. It's a bleak landscape. Almost lunar. At first glance, it's the kind of land you'd never want to explore, much less settle down in. In the small world of geologists, though, the region is well-known. In fact, you might even say it's the single most important patch of undeveloped, unloved, and desolate looking land in America. But you'd never guess this particular corner

of the Great American Desert may play an integral role in
America's strategic future just by looking at it. You'd never
guess that the whole stretch of brown, red, and orange land
contains enough recoverable oil and gas to make you forget
about the Middle East for the rest of time.

Oil shale deposits in the eastern United States are extensive in
Kentucky, Ohio, Indiana, and to a lesser degree in some other states,
but these deposits do not hold any promise in the near-term or medium-
range future.

There are problems with oil shale to be sure, but the challenges
can be ably managed. I am convinced that avoiding other problems and
successfully developing a U.S. oil shale industry requires the public
and policymakers to sidestep Big Oil. What has it done to develop our
vast unconventional oil resources or to make our country petroleum
independent? The U.S. government should encourage investment syn-
dicates to provide the money and the talent to secure properties and
facilities necessary to produce oil from oil shale. It is difficult to trust
Big Oil, and small companies move faster and are more innovative.
Even if Big Oil does begin to exploit oil shale, the U.S. government
should make it easy for other investors to actively enter this business.
Of course, one would extract oil from the most concentrated deposits first. If those mining leases are tied up, then leaseholders should be
required to release them unless the holders commit to exploitation over
the next 12 months.

Much of the following information comes from various reports
listed in the references section of this book, but the principal source
is *Strategic Significance of America's Oil Shale Resource, volume II: Oil Shale
Resources Technology and Economies.* Another valuable source is Robert
Loucks' *Shale Oil: Tapping the Treasure.*

Most of the extensive work on American oil shale was discontin-
ued when oil was less expensive. However, everybody in government
and in the oil industry must have known about the U.S. and worldwide oil supply becoming dangerously depleted in a relatively short

time—or they darn well should have known. They also should know the time it takes to bring oil from unconventional sources into production. The purely economic excuse to end research and development of oil shale sources was, under the circumstances, incredibly stupid or naïve in my view, because simple supply and demand tells us that oil prices will increase at an accelerating rate as oil becomes more scarce.

## Oil Shale History

Using oil shale as a fuel dates to around 1840. A fuel shortage during World War I accelerated the exploitation of this resource in the United States. In Estonia, mining began in 1918 and continues today. Oil shale fuels 2 large Estonian power stations: A 1400-megawatt plant opened in 1965, and a 1600-megawatt power station went online in 1973. These plants could provide enough energy for over 2 million U.S. households. Eventually a nuclear plant reduced the need for Estonian shale since nuclear power is cheaper and much less polluting. Brazil, China, and Russia have also been producing some oil from oil shale for a long time on a limited basis. The relatively low price of conventional oil in recent decades has discouraged the more expensive recovery of oil from oil shale.

For the last century experts have known the vast extent of U.S. oil shale, more than 7 times the oil reserves of Saudi Arabia. How lucky can we get? Right here in the good ol' USA, we have oil reserves in the ground to become oil-energy independent or at least enough oil to last through the planned 30-year transition—a bridge—to other, cleaner, renewable energy sources.

It has been nearly 20 years since anything has been done to develop U.S. oil shale deposits. In that time technology has advanced, the global economic, political, and market conditions have changed, and the regulatory landscape has matured.

## Where Is the Oil Shale?

Oil shale, deposited across millions of square miles, is known to exist in at least 100 major deposits in 38 countries. Some deposits can yield up to 100 barrels of oil per ton of oil shale. Obviously the key question is how much can be economically recovered. I will discuss only those deposits containing 25–100 gallons of oil per ton of shale. On this basis world reserves could be as high as 8 times all the conventional oil ever discovered.

As a worldwide average, it would take about 1300 square miles to produce 1 billion barrels of oil from oil shale. By comparison, the Green River area of the western United States would require only 14 square miles to produce 1 billion barrels of oil. At present, less than 1 percent of all oil shale deposits are economically feasible for oil production. The U.S. Energy Information Administration estimates the world supply of oil shale contains 2.9 trillion barrels of recoverable oil. Reserves of over 750 billion barrels—that is, 25 percent of recoverable oil from oil shale—sit in the United States. In contrast, the United States has fewer than 20 billion barrels of conventional oil in reserve. The U.S. Office of Naval Petroleum and Oil Shale offers different numbers. It estimates oil shale worldwide contains 1.6 trillion barrels of recoverable oil, of which 1.1 trillion barrels are in the United States. Both sets of numbers are supportable depending on the assumptions made. Bottom line: There is a lot of oil in oil shale around the world, and the United States has most of the good stuff (see *Shale Oil* by Robert Loucks).

### Muddle or Manage?

In *Shale Oil* Robert Loucks quotes Dr. Armand Hammer, one of the country's most innovative and notorious industrial leaders, a man who invested over $100 million in recovery and processing of oil shale through his firm Occidental Petroleum. Hammer's statement from 1979, when he was over 80 years old, remains relevant today.

> The current and future availability of crude oil is the nation's most pressing energy problem. No other fuel source or known technology can be expected to power our cars, trucks, tractors, planes and ships, nor our machines of defense, within this century.

We must develop alternative liquid fuels. One such alternative fuel—shale oil—is ready now, and could be providing us with significant quantities of liquid fuels by the mid-1980s. This nation has vast quantities of oil shale. We now have the technology for recovering oil from this domestic resource. What we have so far lacked is the determination to commit the resources and provide the incentives necessary to develop a national shale oil capability.

Since the Oil Embargo of 1973, this nation has done little to expand domestic production of liquid fuels or to develop alternative energy sources. Charles Schultz, Chairman of the President's Joint Council of Economic Advisors, stated this most directly in his testimony before the Joint Economic Committee, Subcommittee on Energy, on April 25, 1979: "our controls on oil discourage the development and production of alternative energy sources. Every barrel of oil equivalent that is produced from non-oil sources, like synthetic fuels, saves the nation some $17-$18 in reduced oil import costs. [How prices have changed.] But we force producers of alternatives to sell into a market where they just compete with oil, whose price is controlled at the lower average price. Thus we discourage investment in such alternative sources, at a direct loss to the nation. *Incredibly, under the current control system, we pay OPEC more for oil than we are willing to pay Americans who produce oil substitutes. It would be hard to design a system more carefully calculated to encourage oil imports and slow down the development of alternative domestic sources"* [emphasis in original].

A number of technologies hold promise for the future—solar, geothermal, nuclear fission and fusion. But none of these is capable of making major contributions to supplement or replace our dwindling petroleum supplies during the next ten to twenty years. You cannot run an automobile on any of these energy forms.

Oil from the rich oil shale deposits in the Rocky Mountains can provide significant quantities of domestic liquid fuels at a reasonable price over the near term. The technology for the large-scale production of shale oil has been demonstrated. We know how to do it. And we believe we can do it in an economic and environmentally acceptable manner. Over 500,000 barrels of oil from shale have been recovered by other private companies and us. This shale oil is compatible with our existing fuels, as produced shale oil without upgrading has been burned directly in utility boilers. Shale oil has been refined to provide gasoline and has been used successfully as a jet aircraft fuel by the Navy.

In testimony before the Congress, I have outlined a plan to foster the development of a two-million-barrel-per-day shale oil industry by the year 1990. [About 10 percent of our current needs.] I have offered the government free license for defense purposes to Occidental's Modified In Situ oil shale process, and we stand ready to license other commercial firms for a reasonable fee.

The Oil Embargo of 1973 warned us of the danger we face of having the major supply of our imported oil cut off at any time. We know that oil worldwide is a depleting asset and will decline significantly over the next 10 to 20 years. Six precious years have passed since the Embargo. The need is here, now, today, to begin the very large-scale

America never really had a coherent energy policy. The United States muddled through as national leaders let Big Oil and Big Coal call the shots.

production of oil substitutes if this nation's economy is to survive. In this document, I have set forth a blueprint for building energy independence from imported oil. Remember, you cannot drill a dry-hole in shale; the oil is there. If we can put an American on the moon, surely we can get this oil out in an environmentally acceptable manner at a reasonable cost. I am confident we can do the job, and I am proud that Occidental is leading the way.

Hammer wasn't the only critic and seer. Many citizens officially voiced their concern. I too joined that chorus in 1973, when I wrote a report entitled *The Oil Crisis in Perspective* for the U.S. Congress. America never really had a coherent energy policy. The United States muddled through as national leaders let Big Oil and Big Coal call the shots. We need a national renewable energy plan—now. You are holding a plan—this book—in your hands. Muddling through won't work anymore, if it ever did.

## Technology

Energy firms typically extract oil from oil shale with either of 2 processes. It can be mined as in conventional ore-mining operations and then processed in a plant, or the oil can be recovered by fracturing the shale underground and then heating the shale in the ground to extract the oil with a conventional oil rig. Plant processing is similar to the processing of oil sands in Canada. Both extraction processes are economically feasible today, and costs will drop in future.

Mining technologies continue to advance. I strongly recommend that U.S. firms immediately test the effectiveness of the Alberta Taciuk Process (ATP)—developed in Canada specifically for mining Canadian oil sands—for recovering oil from U.S. western shale deposits. Since oil shale deposits can vary in specific characteristics, the same process may not work in all cases. To my knowledge ATP has not been tested for U.S. western oil shale.

Mining costs are a minor economic factor, since direct mining costs about $1 per ton, which translates to about 4¢ per barrel of oil. Restoration procedures for depleted open-pit mines are well known and do not present an economic or environmental problem.

Shell Oil has patented a process for heating shale in the ground for up to 4 years to release the oil, which is then pumped out of the ground in the conventional manner. This process promises to produce fuels from oil shale in a manner more environmentally sound and economical than the conventional surface process. However, this process will be too late to contribute much to the bridge.

Experts conclude that the current state of shale oil production, mining, processing, and technologies are sufficiently advanced to support the immediate implementation of a new generation of oil shale projects that could go on-stream in 5 years. A dedicated nation could do it sooner. Experts also tell us that the development and commercialization of a domestic oil shale industry producing 2 million barrels per day by 2020 is in reach. Such production would cover about 10 percent of our present oil needs and about 7 percent of our needs in 2030. Not good enough, but a start.

**Costs**

The economics of oil shale are confusing since many conflicting numbers circulate. Production figures range from $25–40 per barrel, depending upon the depth of the oil, mining and processing methods, and concentration of oil per ton of shale. However, production costs of less than $40 per barrel, the highest cost I've seen, make oil from oil shale not only competitive with conventional oil, but also an incredible investment opportunity. The $40 figure includes all costs, including mining, transporting, refining, disposing of depleted shale, cleaning up the environment, and using approximately 40 percent of the produced energy during the processing.

What is the basis for such optimistic projections and positive statements regarding oil shale? The answer is the experience with oil sands that are similarly mined and processed in Canada. There is no doubt about it—recovery of oil from oil shale is cost effective today and will be even more economical over time.

Consider Australia's oil shale program. Australian firms calculated in U.S. dollars that *the cost of recovering oil from oil shale is about the same as the cost of offshore oil, and with a lot less risk.*

## Environmental Issues

Ecological concerns over the use of oil shale have been thoroughly researched and are quite manageable. The environmental issues arising from the mining and processing of oil shale are the same issues that occur with many other mining and processing operations. Disposal of depleted shale and its remediation does not present any particular technical challenge, and the cost of disposal is usually included in pricing plans and implementation proposals.

Western oil shale is located in a relatively concentrated land area, but mining and processing oil shale could injure the environment. The region is largely rural and is remote from larger population centers. The small towns presently make their livelihoods from producing oil and gas, ranching, and agriculture. However, oil shale development

> The public pays some environmental price for deploying every known energy source—even for wind and solar energy. The best we can do is to choose wisely between the alternatives. The worst choice is coal. The next worse choice is oil.

is consistent with historic activities and is an extension of what people and firms in these areas are already doing. As mentioned earlier, the sites of greatest interest are close to current oil and gas production, where roads and utilities are pretty well developed. Recovered oil could be piped through existing pipelines as current conventional oil production continues to decline.

To be sure there will be temporary environmental damage. For those people who want and enjoy relatively inexpensive energy, it is naïve to think we don't have to pay some environmental price. The public pays some environmental price for deploying every known energy source—even for wind and solar energy. The best we can do is to choose wisely between the alternatives. The worst choice is coal. The next worse choice is oil.

With today's knowledge and technology, environmental damage can be minimized. Also as far as oil shale is concerned, the United States has a good reference point in Canada's oil sand industry, which has dealt with environmental issues and set aside funds for eventual remediation.

## Oil Shale Is the Best Bridge

There is no reason we can't get started right away. As early as 1978 the U.S. Department of Energy concluded that the development of domestic oil shale was technically feasible and was ready for the next steps toward aggressive commercialization. Only political leadership was lacking. Relatively low oil prices in the 1980s lulled us to sleep, but even then our political leaders were warned of pending price hikes for oil based on the dwindling supply.

Developing oil shale builds upon the advantage of a pretty good infrastructure of existing roads and pipelines originally built to transport oil, gas, and other materials. Since oil and natural gas are indigenous to the area, the gas could be used in the recovery of oil from shale. There is also sufficient electrical energy in the region to support the early stages of development. Future electrical needs could and should be provided by clean nuclear power.

Large quantities of water will be required to develop a large-scale oil shale industry. Most of the water will be required for reclaiming the land and supporting the population and economic growth associated with this new industry. Fortunately, there is enough water to support production of 2 million barrels of oil per day from oil shale. The country should begin immediately to engineer and build a processing plant and necessary town facilities to support a growing oil shale industry. These activities should be done simultaneously, since we don't have time to do them consecutively.

Although Big Oil has no obligation to enhance the long-term energy security of the United States, the public similarly has no obligation to provide it subsidies. It seems oil companies find it easier and more profitable to invest in lobbyists, political favors, and ANWR drilling than to develop our vast oil shale reserves. To be sure, oil companies will continue to grip the levers of governmental influence they control and will continue to foist misleading statements upon an inattentive and unsuspecting public.

## Other Considerations

If the United States can produce 2 million barrels per day with current knowledge, then we can and should produce at least 3–5 times more as knowledge and experience develop, and the United States should aggressively supplement this production with cellulosic ethanol and algae biodiesel. This amount, along with some modest imports, should be enough to support the proposed 30-year transition plan. No matter how fast we transition to clean, renewable energy, we will need liquid fuels as a bridge to a clean-energy future.

My understanding is that oil shale plants will be required to obtain over 50 permits and approvals from all levels of government. These permits can take years to obtain. Too many and too long. Some permits are based on EPA-required studies that cannot resolve the controversies at issue with present test procedures, yet long delays result from these inadequacies. Let's hope that governments will streamline these procedure in light of their importance. Once permits are issued and developers comply with the terms of the permit and with all other laws and regulations, litigation to delay and harass this industry should be greatly restricted. Should litigation be permitted, a fast-track process should exist.

## BOTTOM LINE

▶ The bridge for a 30-year transition to all renewables will be difficult, but manageable. A bridge for a 50-year transition will be *very* difficult, and consequently the world will probably be unable to avoid *extreme* economic difficulties.

▶ The United States should not count on Alaskan or offshore oil to help us much in transition. There isn't enough there.

▶ The United States must begin immediately to exploit oil shale opportunities. Inactive leases should be terminated. The main criteria for authorizing permits should be how quickly the project can begin producing.

▶ The U.S. government must do everything possible to make it easier for an oil shale industry to flourish. Permitting processes should be streamlined and regulations simplified and liberalized, while protecting all stakeholders.

▶ Produce 2 million barrels of oil per day within 3 years and 10 million barrels daily within 5 years. Anything less may frustrate our march toward clean, renewable energy independence.

▶ If necessary (though I doubt it will be needed), the government should guarantee a minimum price to protect risk investments.

▶ I believe that recovering oil from oil shale as a bridge is not just an option, but is an absolute necessity. Therefore, we must begin to exploit this resource immediately.

Planned oil shale production should anticipate the needs of other countries wanting to bridge themselves to a renewable energy future.

## Chapter 16

# ENERGY INDEPENDENCE BY 2040

At last, at last ... the good news chapter. In short, the United States can afford clean, renewable energy and energy independence in the relatively near future.

I am *extremely* happy to report that the transition to all-renewable, clean energy sources, if done over the next 30 years, will cost the American people *nothing*. The direct *cash benefits* will actually exceed the *cash outlays*. I tell you how in this chapter. However, delays will be extremely expensive, since the costs of a transitional "bridge" rise dramatically with time. Further, if the United States makes the transition in 30 years—and I know we can with proper leadership—then we should be able to stave off a deep, broad, worldwide, energy-related economic depression. I am excited about the future, and I am anxious for you to consider the following.

Thirty years is the magic number, the realistic window of opportunity, because within this timeframe the necessary changes are achievable. Twenty years is too little time for the magnitude of change required. Forty years is too long, because it brings the United States and the world too close to catastrophe. Recall that with business as usual, conventional oil will be essentially gone in about 30 years. The more quickly the United States embarks on this transition, the more we extend the reserves of oil at hand, and the less we will have to rely on unconventional sources.

The costs of nuclear energy are competitive with the costs of generating electricity from coal, oil, and natural gas. Most studies also show that nuclear power is considerably less expensive than wind power and very much less expensive than solar power, except in some specific applications. In any case it is pointless to argue over these differences, because nuclear power is clean, the least expensive, and also *inevitable*. Regardless of the cost, wind power and solar energy make sense in some applications, such as using solar power for meeting summer peakload needs or for where electric grids do not make sense, as in Indonesia.

Keep in mind that when analysts assess the cost competitiveness of nuclear energy, they include the costs of decommissioning facilities and disposing of waste. However, when analysts calculate the costs of generating electricity from non-nuclear sources, they rarely calculate such "external" costs. That's not comparing apples to apples, so is confusing. Actually, it is worse; it is a deceptive comparison. If one considers the external costs for coal, oil, and natural gas—that is, the costs for society, health, and the environment—then nuclear energy would cost many times less than energy from coal and considerably less than energy from natural gas. And this calculation does not include the consequences of possible global warming. The same comparison would show that energy from wind, solar, and other renewable sources costs less than the continued use of fossil fuels.

Procrastination is quickly becoming our most formidable obstacle to clean power, energy independence, and a reasonable economic future.

Arguments about what causes global warming or tipping points do not really matter since we have learned there are many other reasons that compel us to abandon fossil fuels. It would be idiotic to bet our children's or grandchildren's lives on the outcome of such debates over climate change. Quite simply we must get off the fossil fuel disaster train. When we do, the possibility of global warming and other impending environmental disasters will disappear. It is time to stop being fossil foolish and become energy wise.

Imagine today's costs to society, health, and environment from the use of fossil fuels. Now grab your calculator. Considering the present consumption rate of fossil fuels and the anticipated growth of 60 percent in the world's population over the next 50 years, the cost of fossil fuels in the future will be dramatically higher than today. I believe fossil fuels will be rationed in the near future, even if the United States immediately begins producing electricity using more nuclear, wind, and solar energy and even if automakers accelerate the development and deployment of hybrid plug-ins and all-electric vehicles. Procrastination is quickly becoming our most formidable obstacle to clean power, energy independence, and a reasonable economic future.

We no longer have the gift of time. U.S. leaders in Washington don't seem to have a clue about the urgency. Nor do the 2008 candidates for president appreciate the urgency. They won't unless we let them know we want action—now. We have the necessary technology and the talent, but who will lead us?

> The costs of nuclear energy are competitive with the costs of generating electricity from coal, oil, and natural gas.

My main concern is that we will run out of time to get clean renewable energy sources in place before there is great disruption in global and national economies and before fossil-fuel pollution causes accelerating, irreversible damage to the earth. And let's not forget a few oil wars and oil spills along the way. It takes approximately 5 years to design and clear the regulatory process and a minimum of 5 years to construct a large nuclear plant. Count on 10 years. Too long, so shorten it. Japan took a little over 4 years to design and build a 1300-megawatt-electric reactor, which began operation in 1997. The United States must "fast track" the process.

## USE of NUCLEAR POWER PROVIDES MANY BENEFITS

When the United States transitions to nuclear energy, we will have all the clean electrical energy the country will ever need. Here are 11 more advantages to developing and using nuclear energy:

1. We save the fossil fuel we have for future generations to use as feed stock for the chemical, pharmaceutical, and plastics industries.

2. We need not send our young people to die in foreign lands to secure oil or natural gas to keep our economy afloat. *Each American consumes about 24 times more energy than a person in China, about 45 times more than a person in India, and about 2 times more energy than a person in the European Union.* Most of the world is starting to figure out something is wrong with this picture. Sooner or later the leaders and citizens of other nations will want to correct this marked imbalance since their people want to share in the good life that only abundant energy can bring as a reward for their labors. Seeds of conflict?

3. We eliminate the pollution caused by fossil fuels. Pollution kills well over 2 million people per year worldwide and 50,000 in the United States. This pollution also destroys or stunts forests and wildlife, poisons lakes and people with mercury, and changes the ecology of the oceans.

4. We would no longer have to worry about the so-called tipping point at which global warming spirals out of control and destroys human life on earth, as some predict.

5. We could abate or eliminate poverty worldwide. Education would follow, then better population control.

6. We can alleviate the dire water shortages anticipated for much of the world. With abundant, inexpensive energy, potable water can be produced from sea water or recycled from other sources.

**7.** We eliminate one of the major reasons for conflict—imbalances of energy, water, and other resources.

**8.** We will not have to worry about escalating fuel costs. The price of uranium can rise 100-fold and fuel costs for fast neutron reactors would still be negligible. Of course, the fuel costs for wind and solar are zero.

**9.** We will have inexpensive, clean, electrical energy to support an all-electric automotive fleet. The result will be dramatically lower automobile-operating costs. Here again, consumers would never have to worry about fuel costs increasing.

**10.** We eliminate the present problem of nuclear waste. This "waste" would be used for fuel in fast neutron reactors. In turn, the waste from fast neutron reactors would have to be stored for just 300–500 years instead of over 10,000 years.

**11.** Economic benefits: Balance of payments greatly improved, value of the dollar restored, significant jobs created, and reduced cost of living for all.

Solar and wind energy costs are discussed in their respective chapters. We must now discuss the cost of nuclear energy.

## Calculating Costs

The costs of nuclear energy have often been bandied about with no accurate basis, even lied about, so I feel it important to give you some sense of the real cost. Since fuel costs for nuclear plants are a minor portion of total generating costs, capital costs are the greatest financial concern. At present, capital costs represent approximately 55 percent of the total operating costs for a fast neutron reactor and about 45 percent of the total operating costs for a coal-burning plant.

Analysts and experts draw wildly different conclusions, depending on their assumptions, about the economic viability of nuclear energy. I wondered about such divergent views, so conducted my own economic analysis for a 1000-megawatt-electric fast neutron reactor with internal pyroprocessing. I came up with a total cost of approximately 4.2¢ per kilowatt hour. Although I have done many such analyses in the past, in this case I was assisted by several others who helped me with the assumptions and the final analysis. I wanted several sets of eyes on the calculations. A cost of 4.2¢ per kilowatt hour is fairly consistent with other studies, even though some put this cost as high as 8¢–10¢ per kilowatt hour. However, I've not seen any quantified justification for such high numbers, and I'm quite certain there aren't any.

Here are my assumptions for the cost calculation:

- Plant cost @ $1500 per installed kilowatt of output (Argonne National Laboratory estimate).
- Plant life @ 40 years.
- Cost of money @ 8 percent.
- Plant operating factor @ 85 percent.
- Construction time @ 7 years.
- Security, operating, waste disposal, and maintenance costs @ 1.65¢ per kilowatt hour.
- Decommissioning cost @ 0.1¢ per kilowatt hour.
- Fuel cost and fuel preparation costs negligible.

The basic economic computation is not complicated. Of course, costs will vary with different assumptions. For example, the actual operating life of a nuclear plant could be as much as 60 years, thereby lowering the cost. Also, the cost of money (interest) could be more or less. Further, when fast neutron reactors are used instead of light-water reactors, significant savings result from the reduced cost of waste disposal.

Let's look at some other data and studies.

## Economic Specifics

According to a report published in 2005 by the Organization for Economic Co-operation and Development (OECD)—an international organization of 30 developed countries that accept the principles of representative government and a free market economy—nuclear-generated electricity from a conventional light-water reactor costs 5.28¢ per kilowatt hour, of which 3¢ represents capital costs. Coal-generated electricity costs 4.70¢ per kilowatt hour, with about 2¢ representing capital costs. Electricity generated by natural gas costs 4.56¢ per kilowatt hour.

The surcharge of $50 per ton of coal burned (called for in this chapter) would add approximately 1¢ per kilowatt hour to the cost of electricity from a coal plant. Coal-generated electricity costs would then rise to 5.7¢ per kilowatt hour. According to Robert Socolow and Stephen Pacala in *Scientific American*, a tax of $200 per ton of carbon emissions—that is, a tax on the carbon emitted from burning coal—has been proposed in the United States, yet even this would not be enough to compensate for the environmental damage.

Since numbers can vary depending on authors' assumptions and the year published, I decided to study several sources—such as OECD, World Nuclear Association, Nuclear Energy Association, the Uranium Institute, U.S. Energy Information Administration, and others. *The main conclusion* is that the numbers are similar and fairly consistent when assumptions are normalized. By tweaking a variable—depreciate over 30 years versus 60?—you obviously get different numbers, but they can be factored for comparison.

The OECD projects the costs for generating electricity for various countries in 2010. The costs appear below as "U.S. cents per kilowatt hour" (see Figure 16.1). OECD anticipates that nuclear energy costs in 2010 will be lower than fossil-fuel-generated electricity.

## Figure 16.1. Comparative Cost Projections for Generating Electricity in 2010

|  | Nuclear | Coal | Gas |
|---|---|---|---|
| Finland | 2.76 | 3.64 | -- |
| France | 2.54 | 3.33 | 3.92 |
| Germany | 2.86 | 3.52 | 4.90 |
| Switzerland | 2.88 | -- | 4.36 |
| Netherlands | 3.58 | -- | 6.04 |
| Czech Rep | 2.30 | 2.94 | 4.97 |
| Slovakia | 3.13 | 4.78 | 5.59 |
| Romania | 3.06 | 4.55 | -- |
| Japan | 4.80 | 4.95 | 5.21 |
| Korea | 2.34 | 2.16 | 4.65 |
| USA | 3.01 | 2.71 | 4.67 |
| Canada | 2.60 | 3.11 | 4.00 |

Costs calculated assuming: US 2003 cents per kilowatt hour, discount rate of 5 percent, and 40-year lifetime of electricity plant operating at 85 percent capacity factor.
Source: OECD/IEA NEA 2005.

## University of Chicago Study

"The Economic Future of Nuclear Power," a study conducted at the University of Chicago, is probably the most complete and in-depth study of the current costs of energy. Professor George S. Tolley, professor emeritus at the University of Chicago, and Donald W. Jones, vice president of RCF Economic and Financial Consulting, Inc., co-led the study. They were assisted by 17 graduate students and advanced undergraduate students with varying, relevant academic backgrounds. In addition, over 25 highly credentialed and respected scientists and economists contributed their considerable knowledge to this study. In short, this study was conducted by an eminently qualified group.

This study goes out of its way, where many other studies didn't or couldn't, to compare like costs, to compare apples to apples so to speak. The study includes both light-water reactors and fast neutron reactors (called integrated fast reactors by the researchers). The study also covers all operating and capital costs, even for early, new, and pro-totype nuclear plants. Prototypes, of course, cost more and are called first-of-a-kind-engineering plants; they are of little interest here. I concentrate only on plants that operate beyond the testing stages, have emerged from the operational learning curve, are of stable design, and have achieved operating "maturity." Also, I focus only on *busbar costs*, that is the cost of electricity delivered to the grid. The study generally reports costs per megawatt hour, but I converted this to the cost per kilowatt hour, since this is the unit used to bill you for electricity.

The researchers determined the cost of electricity generated from a mature light-water reactor. If the plant was built in 5 years and operated at 85 percent of capacity, then the cost would be 4.7¢ per kilowatt hour. This includes *all* costs, including long-term storage of the nuclear waste and the decommissioning of the plant at the end of its life cycle. If the plant took 7 years to build, then costs would increase to 5.3¢ per kilowatt hour. However, if the capacity factor at the same plant increased to 90 percent, then the cost would drop to 4.4¢ per kilowatt hour. Although other variables affect the conclusions, these figures illustrate how costs vary with even modest changes to a few variables.

In comparison, a pulverized-coal plant produces electricity for 3.3¢–4.1¢ per kilowatt hour. A natural gas plant generates electricity for 3.5¢–4.5¢ per kilowatt hour. These figures assume relatively stable prices for coal and gas. Based on use patterns, population growth, and dwindling supplies, the assumption of stable prices makes no sense, since we all know that prices are escalating and will continue to escalate. For example, the study anticipated the natural gas prices in 2005 to be $3.61 per 1000 cubic feet. In fact, it was about $7.50 per 1000 cubic feet. Coal prices have also escalated from projections.

Uranium prices, projected to be $15 per pound through 2006, actually rose to $72 per pound ($160 per kilogram) in summer 2007. This increase has little effect on the overall price of electricity from a light-water reactor and no effect whatsoever for a fast neutron reactor.

The fuel cost for a light-water reactor is less than 0.2¢ per kilowatt hour, and the fuel cost for fast neutron reactors is almost zero, less than one-hundredth of a cent per kilowatt hour. Even if uranium prices increase 100-fold, fuel costs for a fast neutron reactor remain negligible.

Other variables affect the price of electricity. In 2004, the most economical power plants fired by natural gas and coal begin with about the same costs and produce electricity at 3.5¢ per kilowatt hour. The costs necessary to reduce emissions raise the price of electricity from a gas plant to 5.2¢–6.4¢ per kilowatt hour; for coal plants, to 7.1¢–10¢ per kilowatt hour. A carbon tax would further raise costs.

The most economic and mature nuclear plant, an advanced light-water reactor, produces electricity for 4.7¢ per kilowatt hour. If the U.S. government would guarantee the capital investment, provide accelerated depreciation, and offer an investment tax credit of 20 percent in a show of support, then the cost of nuclear energy would fall by approximately 1.06¢ per kilowatt hour, thereby lowering the price to 3.64¢ per kilowatt hour. If a proposed production tax credit were also added, then the price would drop 2.1¢ per kilowatt hour. In the interest of time, I do not recommend any government subsidies; I recommend only outright capital grants to build energy infrastucture. The government should become an aggressive facilitator, but not a grantor of subsidies, which are often arbitrary and cause delays.

> The most economic and mature nuclear plant, an advanced light-water reactor, produces electricity for 4.7¢ per kilowatt hour.

Some uncertainty surrounds costs of fast neutron reactors, but all indications suggest these reactors will be just as economical, or more so, than light-water reactors. The study from the University of Chicago does not show this, since it projects a price of 5.7¢ per kilowatt hour, but the study's researchers did not give credit for the more efficient use of fuel or for the much lower costs of waste disposal.

The paragraphs immediately above are a bit tedious, but I thought them important to illustrate costs and factors affecting costs.

Costs for wind and solar energy are thoroughly discussed in their respective chapters. Given the scope and severity of the world's energy problems, small differences of a couple of cents here or there are really meaningless. The bottom line is clear: *We must get off of fossil fuels, and fast neutron reactors are inevitable.* Even adding a penalty of 5¢–10¢ per kilowatt hour for an all-renewable-energy future would be a good buy. Let's end studies and hypothetical planning, and let's move to determined action—now. We need commitment. Once we have it, we will find even more economical and safe ways to produce all forms of renewable energy. I guarantee this. Once in the game, you always get better.

The U.S. government and public must commit to a 30-year transition to a clean renewable energy future. We should also encourage and help the rest of the world to do the same. The benefits cited here apply to all nations.

## ABOUT U.S. ELECTRICITY

The average price Americans pay for electrical energy is about 8.5¢ per kilowatt hour, but in some areas the price can be as high as 10¢–12¢ per kilowatt hour or as low as 7¢ per kilowatt hour.

In 1940 the United States used about 10 percent of its energy resources to produce electricity. Now we use 40 percent of our energy to produce electricity. National security and a modern, healthy economy

> Fast neutron reactors are inevitable.

are impossible without this electricity, which transmits both energy and *information*. We all know how difficult it is for communities to meet basic needs like food, shelter, water, and law enforcement when there is a lack of electricity for even a brief period of time. Remember Hurricane Katrina? Recall riots in major cities during brown outs and black outs?

America has almost 160,000 miles of high-voltage transmission lines. These lines have been neglected in recent years, resulting in grid congestion, which in turn prevents lower-cost electricity from reaching some consumers. It also prevents the diversion of large amounts of electricity to places experiencing a serious shortage or an emergency. The grid needs attention.

## HOW MANY NUCLEAR, WIND, and SOLAR POWER PLANTS DOES the UNITED STATES (and WORLD) NEED and WHAT WILL THEY COST?

The following numbers are the result of my own broad-brush calculations of electricity needs and costs for the world and for the United States. All calculations and charts are based on electricity produced by standardized, 1000-megawatt-electric nuclear plants, 2-megawatt windmills, or photovoltaic solar cells. I assume these plants will be built over a 30-year time span. I calculate the number of plants needed and their cost for 3 scenarios for both the world and for the United States:

- Generating traditional electricity needs only.

- Generating traditional electricity needs plus electricity for an all-electric transportation system.

- Generating electricity for a fully integrated, all-electric economy in which almost all energy is electric. Obviously this will never happen, but as an exercise, it sets the high-end boundary.

The following figures—Figures 16.2, 16.3, and 16.4—are absolutely essential to my argument in this book. I hope you spend a few moments looking at the details. From this information one can determine the capital costs of any combination of nuclear, wind, and solar power plants. Examples are given.

## Scenario 1: Meeting Traditional Electricity Needs

Notice first that the world needs almost 2 thousand 1000-mega-watt-electric plants by 2040 just to **meet traditional electricity needs** (see Figure 16.2). These plants are in addition to the 441 nuclear plants and other renewable-energy facilities currently operating in the world. Building the necessary plants would cost $100 billion annually for 30 years. [Note: In reading Figure 16.2 (and 16.3 and 16.4 below), be aware that electricity already produced by nuclear energy, hydro-power, and biomass were subtracted from the totals.]

## Figure 16.2. Number and Capital Cost of Non-Polluting Power Plants Required to Generate Traditional Electricity Needs Only

| | | Capital Cost (Nuclear vs. Wind vs. Solar)—Electricity Generation Only | | | | | | | | |
|---|---|---|---|---|---|---|---|---|---|---|
| | | World | | | | U.S. | | | | |
| | | | | Build Over 30 Yrs. | | | | Build Over 30 Yrs. | | |
| | Year | (1) Number Required | (2) $ Cost Trillions | No/Yr Required | $ Cost/Yr Billions | (1) Number Required | (2) $ Cost Trillions | No/Yr Required | $ Cost/Yr Billions |
| Nuclear | 2008 | 1,570 | 2.4 | 52 | 80 | 360 | 0.5 | 12 | 18 |
| | 2038 | 2,000 | 3.0 | 67 | 100 | 470 | 0.7 | 16 | 24 |
| Wind | 2008 | 2,000,000 | 7.2 | 67,000 | 240 | 450,000 | 1.6 | 15,000 | 54 |
| | 2038 | 2,600,000 | 9.4 | 87,000 | 310 | 590,000 | 2.1 | 20,000 | 70 |
| Solar | 2008 | N/A | 24.0 | N/A | 800 | N/A | 5.0 | N/A | 180 |
| | 2038 | N/A | 31.0 | N/A | 1,000 | N/A | 7.0 | N/A | 230 |

(1) Nuclear "Number" is number of 1000-megawatt-electric nuclear generating plants required.
  Wind "Number" is number of 2-megawatt generating plants required.
(2) Capital cost per megawatt installed:
  Nuclear @ $1,500/megawatt-electric
  Wind @ $1,800/megawatt (5,400 megawatt-electric) at 33 percent capacity
  Solar @ $6,000/megawatt (18,000 megawatt-electric) at 33 percent capacity
Source: Created by author from various sources.

## Scenario 2: Meeting Traditional Electricity Needs and the Needs of an All-Electric Transportation System

Next, if the world wants to address both the rapid depletion of oil and the attendant pollution, then we must generate sufficient non-polluting renewable energy to **satisfy traditional needs for electricity and the needs of an all-electric transportation system** (see Figure 16.3). The costs are substantial, but the savings are even greater. For example, if all vehicles were electric, then the United States would decrease oil consumption by 4.9 billion barrels a year (valued at $500 billion per year), improve U.S. balance of payments by $300 billion per year, and reduce $CO_2$ emissions by over 1 billion tons per year. The average cost per mile to drive a car would fall to 1¢–2¢, not including taxes. Compare this to the pre-tax 12¢ per mile cost U.S. most drivers now pay. After the transition the cost to drive a car should decrease by about 10¢ per mile, a huge national savings. This represents a giant step toward clean-energy independence. We can afford to pay for it *and* reduce the living expenses for every American. Other countries would achieve the same savings. Does anyone have a better plan?

### Figure 16.3. Number and Capital Cost of Non-Polluting Power Plants Required to Generate Traditional Electricity Needs Plus Electricity for an All-Electric Transportation Fleet

| | | Electricity Generation + Electricity for Electric Transportation | | | | | | | | |
|---|---|---|---|---|---|---|---|---|---|---|
| | | World | | | | U.S. | | | | |
| | | | | Build Over 30 Yrs | | | | Build Over 30 Yrs | |
| | Year | (1) Number Required | (2) $ Cost Trillions | No/Yr Required | $ Cost/Yr Billions | (1) Number Required | (2) $ Cost Trillions | No/Yr Required | $ Cost/Yr Billions |
| Nuclear | 2008 | 2,800 | 3.8 | 93 | 130 | 650 | 0.9 | 21 | 30 |
| | 2038 | 3,700 | 5.0 | 123 | 170 | 850 | 1.1 | 28 | 37 |
| Wind | 2008 | 3,500,000 | 12.6 | 117,000 | 420 | 810,000 | 2.9 | 27,000 | 97 |
| | 2038 | 4,800,000 | 17.3 | 160,000 | 580 | 1,100,000 | 4.0 | 37,000 | 133 |
| Solar | 2008 | N/A | 42 | N/A | 1400 | N/A | 9.7 | N/A | 320 |
| | 2038 | N/A | 58 | N/A | 1900 | N/A | 13.3 | N/A | 440 |

(1) Nuclear "Number" is number of 1000-megawatt-electric nuclear generating plants required.
  Wind "Number" is number of 2-megawatt generating plants required.

(2) Capital cost per megawatt installed:
  Nuclear @ $1,500/megawatt-electric
  Wind @ $1,800/megawatt (5,400 megawatt-electric) at 33 percent capacity
  Solar @ $6,000/megawatt (18,000 megawatt-electric) at 33 percent capacity

Source: Created by author from various sources.

## Scenario 3: Meeting Electricity Needs for a Fully Integrated All-Electric Economy

Now let's make a pie-in-the-sky assumption. What would be necessary to **generate the electricity needs for a fully integrated all-electric economy**—that is, the electricity necessary to meet all energy requirements for traditional electrical needs, plus transportation, plus all residential and commercial needs, including heating and cooling, and all industrial energy needs? Probably not practical, but such an economy and energy-producing system would take us to a totally clean, energy-independent future, but this example defines the high limit (see Figure 16.4).

## Figure 16.4. Number and Capital Cost of Non-Polluting Power Plants Required for a Fully Electric Economy

| | | Electricity + Transportation + Residential/Industrial = All Electric Economy | | | | | | | |
|---|---|---|---|---|---|---|---|---|---|
| | | **World** | | | | **U.S.** | | | |
| | | | | Build Over 30 Yrs | | | | Build Over 30 Yrs | |
| | Year | (1) Number Required | (2) $ Cost Trillions | No/Yr Required | $ Cost/Yr Billions | (1) Number Required | (2) $ Cost Trillions | No/Yr Required | $ Cost/Yr Billions |
| **Nuclear** | 2007 | 6,500 | 9.8 | 216 | 330 | 1,500 | 2.3 | 50 | 77 |
| | 2037 | 8,400 | 12.6 | 280 | 420 | 1,950 | 2.9 | 65 | 97 |
| **Wind** | 2007 | 8,300,000 | 30 | 275,000 | 1000 | 1,900,000 | 6.8 | 63,000 | 230 |
| | 2037 | 10,400,000 | 37 | 350,000 | 1200 | 2,400,000 | 8.7 | 80,000 | 290 |
| **Solar** | 2007 | N/A | 100 | N/A | 3300 | N/A | 23.0 | N/A | 770 |
| | 2037 | N/A | 123 | N/A | 4000 | N/A | 29.0 | N/A | 1000 |

(1) Nuclear "Number" is number of 1000-megawatt-electric nuclear generating plants required.
  Wind "Number" is number of 2-megawatt generating plants required.

(2) Capital cost per megawatt installed:
  Nuclear @ $1,500/megawatt-electric
  Wind @ $1,800/megawatt (5,400 megawatt-electric) at 33 percent capacity
  Solar @ $6,000/megawatt (18,000 megawatt-electric) at 33 percent capacity

Source: Created by author from various sources.

The favorable economics of all 3 scenarios would permit wealthy nations to assist poorer nations to achieve energy independence as well. At that time everyone will be breathing clean air and drinking clean water.

After the transition the cost to drive a car should decrease by about 10¢ per mile, a huge national savings.

If you want to determine the capital costs of various combinations of nuclear, wind, and solar, then the above charts are all you need. For example, let's consider Scenario 2 and say you believe that by 2040 the United States should produce 15 percent of its electrical and transportation energy from wind, 15 percent from solar, and 70 percent from nuclear sources. Then, following column (2) under the heading "U.S.," the needed capital costs, drawn from Figure 16.3, are:

$$
\begin{aligned}
&\ 0.70 \times\quad \$1.1 \text{ trillion (for nuclear)}\\
+\ &\ 0.15 \times\quad \$4.0 \text{ trillion (for wind)}\\
+\ &\ 0.15 \times\ \$13.3 \text{ trillion (for solar)}
\end{aligned}
$$

$$
\begin{aligned}
=\quad &\$3.4 \text{ trillion TOTAL, or}\\
=\quad &\$112.0 \text{ billion per year for 30 years.}
\end{aligned}
$$

Let's then say the proposed combination calls for contributions of 25 percent from wind, 25 percent from solar, and 50 percent from nuclear. Here are the costs derived from Figure 16.3:

$$
\begin{aligned}
&\ 0.50 \times\quad \$1.1 \text{ trillion (for nuclear)}\\
+\ &\ 0.25 \times\quad \$4.0 \text{ trillion (for wind)}\\
+\ &\ 0.25 \times\ \$13.3 \text{ trillion (for solar)}
\end{aligned}
$$

$$
\begin{aligned}
=\quad &\$4.9 \text{ trillion TOTAL, or}\\
=\quad &\$163.0 \text{ billion per year for 30 years.}
\end{aligned}
$$

The math presented in the tables above is simple and clear. But there is more. For comparison, the United States spends about $500 billion per year for defense (2007); the world spends $1.2 trillion (see Figure 16.5). Does anyone else think these priorities are criminally whacko? Also consider what is being spent in Iraq and Afghanistan to protect U.S. petroleum sources and political interests— probably $1¢–1.5 trillion before U.S. forces leave Iraq. And still the sources and interests will not be secure. If only we could divert some of this money to improve worldwide education and eliminate poverty. A moratorium on worldwide hostilities and military spending for 5 years would take care of all of the world's electrical needs in 2040. More to the point, once the world makes the transition to all-renewable energy sources, there will be fewer reasons for conflict, and maybe money currently spent for national defense could be used to respect and enhance life, not destroy it. Do we want to continue killing each other for scarce fossil fuels, or would we rather build a secure, clean energy future for all?

## Figure 16.5. Population and Military Expenditures, 2008 to 2038

| Country | 2008 Population Millions | | 2038 Population Millions | | 2008 Military Expenditures Billions of Dollars | | 2038 Estimated Military Expenditures (Same % GDP as 2008) Billions of Dollars | |
|---|---|---|---|---|---|---|---|---|
| | Amount | Global % | Amount | Global % | Amount | Global % | Amount | Global % |
| | | | Estimates | | | | Estimates | |
| United States | 300 | 4.6 | 390 | 4.1 | 520 | 43 | 670 | 39 |
| EU | 460 | 7.1 | 480 | 5 | 220 | 18 | 230 | 13 |
| China | 1,300 | 20 | 1,600 | 16.8 | 82 | 6.7 | 180 | 11 |
| India | 1,100 | 17 | 1,700 | 17.9 | 19 | 1.6 | 80 | 5 |
| Russia | 145 | 2.2 | 140 | 1.5 | 18 | 1.5 | 30 | 2 |
| Rest of World | 3,200 | 49 | 5,200 | 54.7 | 340 | 28 | 510 | 30 |
| World | 6,500 | 100 | *9,500 | 100 | 1,200 | 100 | 1,700 | 100 |

*Population projections range from 9 billion to 10.2 billion. GDP = gross domestic product.

Source: 2007 CIA - World Fact Book; 2038 figures calculated and compiled by author from various sources.

## A WISE MAN SPEAKS: DR. DAN MENELEY'S PRESENTATION in 2006

I was impressed by Dr. Dan Meneley's incisive presentation entitled "Transition to Large Scale Nuclear Energy Supply," a strong presentation on why the world must move rapidly to nuclear power delivered from fast neutron reactors. He presented his report at the twenty-seventh annual conference of the Canadian Nuclear Society in 2006. Dr. Meneley is a highly respected and credentialed nuclear scientist. He holds a Ph.D. in reactor science, and he was the chief engineer (now engineer emeritus) at Atomic Energy of Canada Limited (AECL). He is the current president of the Canadian Nuclear Society. The Canadians almost always seem to get it right.

Here is how Dr. Meneley described "Where Do We Stand?" in his presentation.

- World oil production is at or near its historical peak.

- Most production capacity is controlled by national oil companies—and is not part of a market economy.

- The oil demands of China and India are increasing rapidly—they expect to import mainly from OPEC, but the supply is limited, of course.

- Demand increases must be satisfied by new discoveries— oil extracted from oil sands and oil shale may help satisfy the increasing demand.

My calculations and Dr. Meneley's calculations led us to precisely the same conclusion—the immediate need to move decisively toward nuclear energy—though with somewhat different results at the margins of our arguments. Yet our conclusions are close enough to provide mutual confidence in each other's projections. With his approval I gratefully borrow from his presentation.

Dr. Meneley projects that the world must build 6000 1000-megawatt-electric nuclear plants or their equivalents to satisfy the world's projected needs for electricity and transportation over the next 30 years.

(In the following discussion, any reference to a nuclear power plant means a 1000-megawatt-electric nuclear power plant.) He contends, and I agree, that to build 6000 plants over the next 30 years—that is, 200 plants per year—is not trivial, but not that difficult a task for a committed world. In comparison, my calculations actually indicate that by 2040 the world needs to build 3700 1000-megawatt-electric nuclear power plants, or approximately 123 per year, after subtracting the nuclear energy and other renewable energy sources already in place. I've also reduced the number of plants required because of my expectation that many more higher-efficiency electric cars (compared to gasoline-powered cars) will be on the roads.

You may find interesting some excerpts from Meneley's speech.

> Conservation, along with a number of alternate energy supply options, has been studied in great detail for a number of years with limited success. *It has slowly become obvious that nuclear energy is the only resource available today that could take over a large fraction of the world demand for oil and gas, and yet remain neither capacity nor resource limited, that is, to be "inexhaustible" or "renewable."* There is enough accessible uranium to supply the total present-day demands of humanity for at least several thousand years....

> Modern assessments differ little from that described in the International Institute for Applied Systems Analysis (IIASA) study carried out more than 20 years earlier. The main change since the IIASA study is that the needed replacement for fossil fuels is now very urgent. Nuclear energy using uranium offers the only practical answer for filling in a major part of the gap between supply and demand. Even then, the enormous scale of the replacement task cannot be over-emphasized. This is not to belittle contributions of other renewable resources and conservation. The statement is meant only to emphasize the central role of nuclear energy in any sound plan, regardless of what other partial solutions are adopted [emphases added].

## FUTURE ELECTRICITY NEEDS

Fossil fuel prices are much higher today than predicted even a few years ago. While prices will drop from time to time, the trend will rise steadily and more sharply. The Middle East holds 65 percent of the world's oil; a glitch in that region could easily put oil at $200 per barrel. Other fossil-fuel-caused problems are on the way—the perfect storm of economic crises triggered by rising fuel prices; gasoline rationing; mounting pollution and the associated deaths, illness, and environmental harm; climate change; and international resentment and anger. Besides, many people, including the Chinese, Indians, Europeans, and Japanese are deeply upset that Americans consume such a huge percentage of the world's energy and other resources. At the same time we must encourage other countries to embark on their own 30-year plans because the same energy issues that apply to the United States also apply to all other countries. We will all run out of fossil fuels together, and we all share the same pollution problems.

## FINANCIAL EPIPHANY: THIS PLAN IS AFFORDABLE.

*Many believe the transition to clean, renewable, eternal energy will break the bank. Not so.* Read on.

You could argue that my numbers are high or low. If they are high, then we should be happy. However, there is a strong tendency for such analyses to be low. As a precaution I built a 10-percent contingency into all my projections. Even counting this 10-percent addition, my figures are still somewhat lower than Dr. Meneley's numbers. So, for planning purposes, let's really put our finger on the scales and, in addition to the 10-percent built-in contingency, add a hefty 25 percent to all of the numbers, except for the cost of solar power, which I believe will fall significantly over the next 30 years. What do those numbers look like?

> ▶ If the United States builds *fast neutron nuclear reactors* to support all U.S. electrical needs and the electrical

needs of an all-electric transportation fleet, then the capital cost per year for 30 years would be:

- 1.25 (i.e., the additional 25 percent) x $37 billion (from far-right column, second row, Figure 16.3) = $45 billion per year for 30 years.

▷ If the United States satisfies all of its electrical needs and the electrical needs of an all-electric transportation fleet with *wind* (practically impossible), then the capital cost per year for 30 years would be:

- 1.25 (i.e., the additional 25 percent) x $133 billion (from far-right column, fourth row, Figure 16.3) = $165 billion per year for 30 years.

▷ If the United States satisfies all of its needs with *solar power* (also practically impossible), then the capital cost per year for 30 years would be:

- $440 billion per year for 30 years (from far-right column, bottom row, Figure 16.3). Note: As mentioned above, I anticipate the costs for solar power will fall, so I do not add an additional 25 percent "contingency" penalty.

**Note:** For wind energy and solar energy, I assumed a service factor of 33 percent. If the wind was blowing all the time and the sun was shining all the time, the above capital costs could be reduced to one-third the numbers shown.

How do we pay for this completely clean, pollution-free, energy-independent future? *The answer is surprisingly easy and painless, but we must insist that the role of government be limited to facilitating the process.* The allotted funds should not come from the government treasury, but should be held in a separate *dedicated fund* managed by, say, the Department of Energy, which will work with electrical utilities across the country. The process—which may serve as a model for other countries—should be designed to *expedite and facilitate* the building of necessary infrastructures to hold at bay the economic meltdown that would result from delay. *The entire plan must be built around speed of deployment.* More positively, government facilitation can fundamentally assist economic growth and adaptation to new and changing market realities. I cannot over-emphasize the urgency. If you doubt it, please review the chapter on building a "bridge" to a future of renewable energy.

## Where Does the Money Come From?

How does the United States get the money, and where does the money come from?

- Each of us should at least conserve as much energy as possible to buy time—time we will need.

- Levy a temporary surcharge of 50¢ per gallon of gasoline, which will yield about $70 billion per year.

- Charge a temporary surcharge of 2.5¢ per kilowatt hour for electricity, which will yield about $90 billion per year.

- Create a surcharge of $50 per ton of coal burned and a surcharge of $1 per 1000 cubic feet of natural gas burned, which would yield about $80 billion per year.

- Apply a gas-guzzler surcharge of 5 percent of the price of the vehicle for cars getting less than 30 miles per gallon, and a surcharge of 10 percent of the price of the vehicle for those getting less than 20 miles per gallon. Total revenue generated? I hope zero.

Here's an important point: The above should more appropriately be called an *investment* because over time the cash returns to every American will be more than this "investment." The environmental benefits would be even greater. Of course, energy for our children and grandchildren to enjoy will be the greatest benefit of all.

The total revenue produced by the relatively modest surcharges listed above would be $240 billion per year. I call these surcharges "relatively modest," because even with these penalties in place, Americans would still pay *about half* of what citizens in other countries pay for their energy. Remember, if we do nothing, then energy costs will rise far beyond these recommended surcharges. Recall also that Americans pay an average of 8.5¢ per kilowatt hour for electrical energy. Yet, Germans pay 18¢ and Japanese pay 21¢ per kilowatt hour for electricity (see Figure 6.1). Europeans pay almost $8 per gallon for gasoline.

For the year 2040, let's assume a reasonable mix of nuclear (80 percent), wind (10 percent), and solar power (10 percent), per the "solar roadmap." Let's further assume that the entire U.S. transportation fleet is all-electric by 2040. Finally, as before, let's assume a worst-case scenario for the United States by saying that costs are 25 percent higher than predicted by the accompanying charts. Under these assumptions, what will the costs be?

Answer: $110 billion per year for 30 years. Here's the calculation from Figure 16.3: 1.25 (i.e., the additional 25 percent) x [(0.8 x $1.1 trillion for nuclear) + (0.1 x $4 trillion for wind) + (0.1 x $13.3 trillion for solar)] = $3.26 trillion total, or $110 billion per year for 30 years.

## IMPLEMENTATION

Here are 19 things the United States must do now. If not now, then we squander valuable and fleeting time.

1. Surcharges must begin immediately.

2. The pilot plant for a fast neutron reactor must be funded immediately, and the funds must be sufficient to accelerate completion of the project. Start now. Do it in 4 years. A crash program must be initiated to accelerate fuel recycling and fast neutron reactor deployment. The program must be overwhelmed with qualified people and money.

3. Fast neutron reactors must be made the foundation for all future electricity-generating systems. Wind, solar, and other renewable energy sources must be added as appropriate.

4. The latest designs of light-water reactors should be built now until fast neutron reactors are ready for center stage. No more coal-fired power plants should be built, and existing coal-fired power plants should be retired as soon as possible.

5. The proposed Global Nuclear Energy Partnership (GNEP) must be aggressively supported and implemented.

6. All renewable energy research must be aggressively funded to accelerate all programs.

7. Research on batteries must be given high priority and amply funded to accelerate development of advanced, high current density, recyclable batteries.

8. Everything possible must be done to ensure a rapid and critical transition away from the internal combustion engine to hybrid plug-ins and all electric vehicles.

9. Ethanol and biodiesel are needed to help "bridge" us to an all-clean, renewable-energy future. The growth of the cellulosic ethanol and algae biodiesel industries must be accelerated.

10. All monies raised by surcharges should go directly to the U.S. Department of Energy, where a separate fund will be set up to provide the capital for the transition to an all-renewable energy future.

11. A special board will oversee this Department of Energy fund. This board could be made up of, say, 4 representatives from government, 4 from industry, and 4 from the country's technical community (from the country's academic and national laboratories, for example).

12. Funds should be allocated to all parties submitting acceptable proposals in order to build energy facilities aimed at an initial goal-mix of 80 percent nuclear, 10

percent wind, and 10 percent solar. Utilities should receive high priority in the interests of time, because they know best the kind of energy that would best fit their circumstances, such as access to land, water, and the electrical grid. Other companies and individuals can also submit proposals for approval.

13. *In order to expedite this plan, monies should be given directly to utilities and others building the new energy infrastructure,* with the only requirement being that the capital costs not be included in subsequent billing to energy customers. This would lower the cost for electrical energy by about 50 percent once these new plants are built, because capital costs represent 50 percent or more of the cost of electricity.

14. The final arbiter on all projects will be the special board cited in Point 11 immediately above. All reasonable regulations must be respected and followed. However, there are many regulations that regulators will tell you are nonsense and arbitrary. These should be investigated and eliminated as appropriate.

15. **This is a "war" that requires a "wartime footing."** As in war and as with wartime decisions, the board must be held harmless from lawsuits. This entire effort is analogous to war, and it must be "waged" to prevent more Americans and others from getting killed in resource wars. Without this protection from lawsuits, the people we would want to serve would likely decline. Some honest mistakes will be made, as they are in any war effort. The alternative is to do nothing while the country is tied up in legal knots. This cannot be tolerated.

16. The government should be extremely active in enabling this process in every way possible, including encouraging the cross-licensing of patents in the interest of rapid deployment of the most cost-effective technologies.

17. The U.S. Department of Energy should monitor and quantify the economic benefits of deploying renewable energy: money saved on healthcare, development of jobs, jobs kept in the country, forest revitalization, balance of payments benefits, and, of course, the reduction of pollutants in water and air, and the effect on temperature change, if any.

18. Incentives should be provided for students wanting to study engineering, particularly nuclear.

19. Finally, if need be, we must drag our political leaders into this process. We need them to lead us in this challenge.

If this transition drags on for 50 years, then the United States and the world would face incredible economic disruptions. In such a case, neither the United States nor the world as a whole will likely have enough oil and natural gas to make a smooth, reasonable transition. Besides, there are no fallback positions or workable alternatives.

## JULY 4, 2040: ENERGY INDEPENDENCE DAY

Let's designate July 4, 2040, as Energy Independence Day. A double celebration. Now to the details.

The surcharges described above can be ended in 20 years. All recommended surcharges would end in 2028, since all necessary funding would be in hand. This could be stretched out over a longer time period, but the costs at the beginning of the transition will be greater than in later years, and there is the very welcome possibility of accelerating the transition. The full transition to clean, renewable energy should be complete by July 4, 2040. *That must be our goal.*

*If we put our considerable national will behind this task, I have no doubt we will succeed because we must.* Implementation of this plan should be invigorating, exciting, and a great boost to the U.S. economy. The countries that transition first will have a tremendous economic advantage over those that delay.

## ECONOMIC EXPECTATION: VERY GOOD NEWS

Figures 16.6 and 16.7 report the overall costs and benefits, the economic bottom line. In short, *the U.S. public invests a maximum of $4.8 trillion and gets back $8.3 trillion in cash.* It's not alchemy. It's savvy planning, energy independence, and a cleaner, healthier world. In addition to the cash payback, the savings derived from a clean environment alone could easily total several trillion dollars. Here are the numbers. Over a 30-year span:

### Figure 16.6. Cost-Benefit Ledger for Energy Independence

| Surcharges Collected through 2028 | | $4.8 trillion |
|---|---|---|
| Transition Capital Costs through 2038 | | |
| Nuclear | 80 percent of total U.S. energy production | $1.1 trillion |
| Wind | 10 percent | $0.5 trillion |
| Solar | 10 percent | $1.7 trillion |
| Transportation | N/A | $0.5 trillion |
| | Total | $3.8 trillion |
| Cash Surplus (for unknown costs [upgrading electric grid?] or refund) $1.0 trillion | | |

### Figure 16.7. Actual Cash Payback to the American People Over the 30-Year Transition from Investments in All-Renewable Energy

| | |
|---|---|
| Automobile fuel savings @ 10¢ per mile net of taxes<br>0.10 x (4.5 trillion miles)/2 (average over 30 years) x 30 (years) | $6.8 trillion |
| Electricity savings (capital charge eliminated)<br>0.025 x ($4 trillion)/2 (average over 30 years) x 30 (years) | $1.5 trillion |
| TOTAL | $8.3 trillion |

## Figure 16.8. Summary of Costs for the 30-Year Bridge/Transition

| Consumers Contribute | $4.8 trillion | See Figure 16.6 |
|---|---|---|
| -Energy Transition Costs | $3.8 trillion | See Figure 16.6 |
| -Surplus for Unknowns (or Refund) | $1.0 trillion | |
| Cash Payback Via Energy Cost Savings | $8.3 trillion | See Figure 16.7 |
| Total Net Benefit | $3.5 trillion | |

For the United States, it is important to look beyond 30 years.

◖❯ U.S. residents will perpetually save $500 billion per year, or about $1300 per person per year in direct energy costs.

◖❯ U.S. residents will enjoy a cleaner environment. The world atmosphere and the oceans will recover only if the rest of the world also converts to clean renewable energy.

◖❯ U.S. residents will have stable energy prices long into the future.

◖❯ The United States will be energy independent, resulting in an economic bonanza.

◖❯ The U.S. balance of payments will dramatically improve, and the U.S. dollar will recover.

◖❯ The U.S. should encourage and help other nations by example to adopt this model—with the same economic benefits. After all, we all live on the same planet.

Let's look at the global consequences, over the same 30 years, if most countries make a similar transition.

◖❯ Oceans would recover, mercury will dissipate, acid rain and toxic gases will disappear—and human-caused global warming will no longer be debated.

◖❯ If the whole world becomes energy independent, then a major cause of conflict, violence, and war would be eliminated.

▷ Last but not least, the legacy given to the world's children will be a reasonably sound economy and ample energy to grow and prosper.

## BOTTOM LINE for the UNITED STATES and the WORLD

There you have it. We can grab the moment and the opportunity, or we can go on with business as usual and apply a multitude of ineffective Band-Aid fixes. Which will it be? Your children and grandchildren want to know.

I am encouraged. Americans typically choose to do the right thing during crises. And this is a crisis. Doing the right thing in this crisis will not cost one American life. In fact, over the proposed 30-year time frame, it will not cost the American people any money either. As the United States helps other nations solve their energy and environmental problems through GNEP and other programs, America would once again assert her leadership role by helping the world win this colossal world energy war—a war at least as serious as any the United States or the world has faced in the past.

Indeed, with diligent effort the United States can transform a colossal problem, a perfect storm of epic proportion and cataclysmic prospects, into a golden opportunity and a glorious future.

I sincerely hope this book stimulates optimism, courage, spirited discussion, and rapid action.

# *Epilogue*

## WE NEED A DEDICATED GOVERNMENT

If the president, your congressman, or your senators do not take an aggressive position in respect to our most dangerous energy problems and they do not make it a priority—SEND THEM HOME. This single issue, if not fixed, puts our nation and other nations at an almost doom's day risk.

While the prospects of solving our energy problem can be invigorating, we need the kind of vision John F. Kennedy instilled in the public when he decided the United States should go to the moon. The nation made a commitment and did it. However, some of us have lost confidence in our leaders doing what is necessary to enable a timely solution. Elected officials have not had the courage to deal with many of the important and difficult issues of the day—poverty, healthcare, campaign finance reform, deficit spending, education, huge negative trade deficits, or pollution—why would the most important issue of all, energy, be any different?

As retired congressman Tim Penny from Minnesota wrote in his book *Common Cents*, "I can't tell you how many times in twelve years in Congress I sat awake at night wondering to myself, 'Why are we doing these things? Why are so many decent and honorable public servants so incapable of acting responsibly on the central issues of the day?'"

Representative Tim Penny goes on to say that bogus bills are sometimes passed:

> It was the first of many examples I witnessed in which one part of Congress engaged in political theatrics purely to satisfy selected special interests. Meanwhile, the general interest of the country was once again ignored as voters witnessed another display of nonsense in the nation's capitol.
>
> The culture of Congress invests vast power in the hands of an elite few lawmakers. It cultivates conformity and punishes originality. It acquires and protects perks that debase the concept of 'public service.' It confuses cowardice with courage, hypocrisy with virtue.
>
> I've seen this culture operate for twelve years. I've seen the ways it encourages legislators to duck decisions. I have seen the way it causes us to invent 'crises.' I have seen lawmakers propose feeble legislative initiatives designed to provide political cover instead of addressing the problem at hand. I have witnessed the waste of time and energy devoted to petty partisan bickering. I have seen good and able legislators ignore their better judgment for fear of angering a constituency or special interest group.

I am depressed when I think about some other fundamental governmental/political issues that hinder our political leaders from doing the job for which they were elected. We need campaign reform, so that the wants of special interest groups can no longer trump good legislation. I've managed and lead a number of companies, and it's always been a fulltime job—how can our president and other national leaders spend so much time on the road raising money? I don't get it. We should adopt a system where politicians have a limited time to campaign and their campaigns should be publicly funded. Our nation can no longer provide special interest groups with the legal means to control our political leaders. I doubt my thoughts on this issue will cause the money-laden business of politics to change, but over time I am hopeful.

I've heard it said that if any of our leaders were to disclose to the American people the dire consequences of not doing enough to solve our energy problems, then the story would be so disturbing that he or she would run the risk of not being elected or reelected.

I believe the opposite is true. The average American citizen has always been able to handle the truth and take the necessary steps, including some sacrifice, for the collective good of our nation. Of late, our leaders have insulted us by deceiving us even on such a grave issue as going to war for—you guessed it—oil. While democracy is our preference, it is not the reason for our keeping a military presence in the Middle East—oil is. Our fine men and women are dying in a foreign land while we and all nations on earth can, with commitment, eliminate the need for energy wars.

The United States decided to invade another country on what seems to be in hindsight bogus data. Our behavior and rhetoric continues to inflame and upset people around the world. We have never been more disliked by even those that at one time had respect for us and our ideals. A study of history and the experience of others should have taught our government more reasonable behavior.

Here is what Winston Churchill had to say about starting any kind of war,

> Never, never, never believe any war will be smooth and easy, or that anyone who embarks on the strange voyage can measure the tides and hurricanes he will encounter. The statesman who yields to war fever must realize that

once the signal is given, he is no longer the master of policy
but the slave of unforeseeable and uncontrollable events.

Rather than fighting for energy and spending prodigious amounts
of money to secure the oil rich Middle East region, we could put ourselves well on our way to energy independence. But no sense "crying
over spilled milk."

Together, despite past mistakes and current difficulties, the people
and countries of the world can solve our shared energy problems. I
think we sometimes forget the power of the idea "We the People."
The people are the government. However, we must make our message
loud and clear. On the issue of energy, we the people must press for
action now.

Some of you may think, however, that our great innovative nation
can solve this problem on the fly—we can't. My fear is that as time
passes, and the problem gets much worse, it will be too late. It will
take decades to exploit alternative, eternal, energy sources. I and others
worry that it could already be too late to make a painless transition.

There are no technical problems keeping us from a final energy
solution. There are powerful special interest groups (fossil fools) however that want to ride the dying horse of fossil fuels because the money
potential is huge in the short run while devastating to all of us in the
long run. Do not be fooled by the rhetoric of vested interest groups.

Some indecision springs from the complexity of the options. But
finally we have to do what we know will solve the energy problem in the
near term (20–30 years) so that we can deploy even better solutions in
the future, when new technology catches up with new opportunities.

Three final, very encouraging points:

- The deployment of clean, eternal energy for our nation
  is actually not very expensive, and it would cause an
  unprecedented economic boom.

- When nations all have the energy they need to grow and
  prosper, they can begin to eliminate poverty.

▶ Most importantly, energy independence for all nations removes a huge barrier to peace and becomes the basis for peaceful interaction and cooperation.

## Further

Energy independence is really not a single issue, since energy weaves its way through just about every aspect of our lives. Our economy and our way of life are dead in the water without adequate, clean, reliable, inexpensive energy. Historians will look back at the early decades of the twenty-first century and know that energy was by far the most important issue of the time. What we do now will have a profound impact on how the rest of history will be written. Will our inaction cause the United States and the world to slip into another dark age shackled by poverty, economic ruin, and war? Or will we choose to take action for a better world?

The future shines brightly. We must find the courage and exercise the will to realize that future and to bring it quickly and securely into being. Shine on.

You have the power. Now promote the right kinds of energy. End fossil foolishness.

# References

Abt Associates, with Computer Sciences Corporation and E.H. Pechan Associates, Inc. (2004) *Power Plant Emissions: Particulate Matter-Related Health Damages and the Benefits of Alternative Emission Reduction Scenarios.* A report prepared for the U.S. Environmental Protection Agency and Clear the Air. [Note: Clear the Air— initiated by The Pew Charitable Trusts through a grant to Pace University—is a joint project of the Clean Air Task Force, the National Campaign Against Dirty Power, the National Environmental Trust, and the U.S. Public Interest Research Group Education Fund]. June. Washington, D.C. and Cambridge, MA: Abt Associates. Available at <http://www.abtassociates.com/reports/Final_Power_Plant_Emissions_June_2004.pdf>. Accessed March 4, 2008.

Adler, Jerry. (2007) "After We Are Gone: If Humans Were Evacuated, the Earth Would Flourish." *Newsweek,* July 23, page 57. Available at <http://findarticles.com/p/articles/mi_kmnew/is_200707/ai_n19398282>. Accessed March 14, 2008.

Appenzeller, Tim, and Dennis R. Dimick. (2004) "Signs From Earth." *National Geographic,* September. Available at <http://findarticles.com/p/articles/mi_hb3343/is_200409/ai_n12951368>. Accessed March 3, 2008.

A.T. Kearney [management consulting firm], Global Business Policy Council.

(2005a) *Foreign Direct Investment Confidence Index.* Volume 8. Available at <http://www.atkearney.com/shared_res/pdf/FDICI_2005.pdf>. Accessed March 10, 2008.

A.T. Kearney [management consulting firm]. (2005b) "Global FDI Recovery Clouded by Savings Glut Overhang, Say Global Executives in New A.T. Kearney Study." News release summary of *Foreign Direct Investment Confidence Index* 2005. December 7. Available at <http://www.atkearney.com/main.taf?p=1,5,1,169>. Accessed March 10, 2008.

BBC News. (2005a) "Air Pollution is Responsible for 310,000 Premature Deaths in Europe Each Year, Research Suggests." February 21. Available at <http://news.bbc.co.uk/2/hi/health/4283295.stm>. Accessed February 29, 2008.

BBC News. (2005b) "The European Union Could Save Up to 161 Billion Euros a Year by Reducing Deaths by Air Pollution, the World Health Organization Has Said." April 15. Available at <http://news.bbc.co.uk/2/hi/science/nature/4444191.stm>. Accessed February 29, 2008.

BBC News. (2004) "Polluted Air from America Could be Damaging the Health of People in Britain, Experts Fear." July 12. Available at <http://news.bbc.co.uk/2/hi/health/3886275.stm>. Accessed February 29, 2008.

Bedard, Patrick. (2005) "The Case for Nuke Cars—It's Called 'Hydrogen'." *Car and Driver*, October. Available at <http://fayfreethinkers.com/forums/viewtopic.php?p=4125&sid=1c19a6771cbe8284a80194bd77a39dff>. Accessed March 13, 2008.

Bidwai, Praful. (2005) "India's Indigestible Oil Gulp." *Frontline*, volume 22, issue 8, March 12–25. Available at <http://www.hindu.com/fline/fl2208/stories/20050422001810800.htm>. Accessed March 10, 2008.

Bjerklie, David. (2006) "Feeling the Heat." *Time*, vol. 167, number 14, April 3. Available at <http://www.time.com/time/magazine/article/0,9171,1176986,00.html>. Accessed March 10, 2008.

Bolger, P.M., and B.A. Schwetz. (2002) "Mercury and Health." *New England Journal of Medicine*, volume 347, number 22, November 28: 1735–1736.

Bourne, Jr., Joel K. (2006) "Fall of the Wild: Our Appetite for Oil Threatens to Devour Alaska North Slope." *National Geographic*, May. Available at <http://www.accessmylibrary.com/coms2/summary_0286-28358659_ITM>. Accessed March 14, 2008.

Bradford, Travis. (2006) *Solar Revolution*. Cambridge, MA: MIT Press.

Bray, Dennis, and Hans von Storch. (2007) *The Perspectives of Climate Scientists on Global Climate Change*. Geesthacht, Germany: GKSS-Forschungszentrum Geesthacht GmbH. Available at <http://dvsun3.gkss.de/BERICHTE/GKSS_Berichte_2007/GKSS_2007_11.pdf>. Accessed March 5, 2008.

Brodkin, Jon. (2005) "Study Blames Diesel for Deaths." *The Metro West Daily News* [Framingham, MA], Daily News Transcripts. February 22. Available at <http://www.dailynewstranscript.com/archive/x1207656976>. Accessed February 29, 2008.

Butler, Rhett A. (2005) "Temperate Forests May Worsen Global Warming, Tropical Forests Fight Higher Temperatures." Available at <http://news.mongabay.com/2005/1205-caldeira.html>. Accessed March 6, 2008.

Caldicott, Helen. (2006) *Nuclear Power is Not the Answer*. New York, NY: The New Press.

Center for Energy and Environment. (2007) "Identifying Effective Biomass Strategies: Quantifying Minnesota's Resources and Evaluating Future Opportunities." Draft version available at <http://www.mncee.org/pdf/biomassreport.pdf>. Final version available at <http://www.mncee.org/public_policy/renewable_energy/biomass/index.php>. Accessed March 11, 2008.

Chernobyl Forum (2003–2005) *Chernobyl's Legacy: Health, Environmental, and Socio-Economic Impacts and Recommendations to the Governments of Belarus, the Russian Federation, and Ukraine*. Second revised version. Geneva, Switzerland: United Nations, International Atomic Energy Agency. Available at <http://www.iaea.org/Publications/Booklets/Chernobyl/chernobyl.pdf>. Accessed March 12, 2008.

Chylek, Petr, Jason E. Box, and Glen Lesins. (2004) "Global Warming and the Greenland Ice Sheet." *Climatic Change*, volume 63, numbers 1–2, March: 201–221. Available at <http://www.springerlink.com/content/r916q94v27030858/>. Accessed March 5, 2008.

Chylek, Petr, M.K. Dubey, and G. Lesins. (2006) "Greenland Warming of 1920–1930 and 1995–2005." *Geophysical Research Letters*, volume 33, L11707. Available at <http://www.agu.org/pubs/crossref/2006/2006GL026510.shtml>. Accessed March 5, 2008.

Cifuentes, Luis, Victor H. Borja-Aburto, Nelson Gouveia, George Thurston, and Devra Lee Davis. (2001) "Climate Change: Hidden Benefits of Greenhouse Gas Mitigation." *Science*, volume 293, August 17: 1257–1259. Available at <http://www.sciencemag.org/cgi/content/summary/293/5533/1257>. Accessed March 14, 2008.

Cohen, Bernard L. (1990) *The Nuclear Energy Option*. New York, NY: Plenum Press.

Conniff, Richard. (2007) "Who's Fueling Whom?: Why the Biofuels Movement Could Run Out of Gas." *Smithsonian*, November. Available at <http://www.smithsonianmag.com/science-nature/presence-biofuel-200711.html>. Accessed March 11, 2008.

Copulos, Milton. (2006) "Testimony of Milton R. Copulos, President, National Defense Council Foundation, Before the

Senate Foreign Relations Committee, March 30, 2006." Congressional testimony. Available at <http://www.senate.gov/~foreign/testimony/2006/CopulosTestimony060330.pdf>. Accessed March 11, 2008.

Copulos, Milton. (2003) "America's Achilles Heel: The Hidden Costs of Imported Oil—A Strategy For Energy Independence." October. Washington, D.C.: National Defense Council Foundation. Available at <http://ndcf.homeip.net/ndcf/energy/NDCF_Hidden_Costs_of_Imported_Oil.pdf>. Accessed March 11, 2008.

Cox News Service. (2004) "Study Projects Death, Illness From Coal Power Plants." June 10. Available at <http://www.nrdc.org/news/newsDetails.asp?nID=1399>. Accessed March 4, 2008.

Csere, Csaba. (2005) "With $3 Gas, We Need Accurate mpg Tests." Car and Driver, December. Available at <http://www.caranddriver.com/features/columns/c_d_columns/with_3_gas_we_need_accurate_mpg_tests_column>. Accessed March 13, 2008.

Davidson, Keay. (2006) "Permafrost Could Speed Up Global Warming: 500 Billion Tons of Extra $CO_2$ Could Be Released, Study Says." San Francisco Chronicle, June 16, page A6. Available at <http://www.sfgate.com/cgi-bin/article.cgi?f=/c/a/2006/06/16/MNGKKJFD5M1.DTL>. Accessed March 5, 2008.

Denning, Dan. (2005) "Stinky Water, Sweet Oil." Daily Reckoning, October 7. Available at <http://www.dailyreckoning.com/Writers/DanDenning.html>. Accessed March 13, 2008.

Diaz, Kevin. (2007) "Federal Sting Obtains Materials for 'Dirty Bombs,' Investigators Reveal; Results of Operation Called for Last Fall by Sen. Norm Coleman." Minneapolis StarTribune, July 12. Available at <http://www.highbeam.com/doc/1G1-166368423.html>. Accessed March 11, 2008.

Domenici, Senator Pete V. (2007) A Brighter Tomorrow: Fulfilling the Promise of Nuclear Energy. New York, NY: Rowman and Littlefield Publishers.

Doxon, Lynn Ellen. (2001) The Alcohol Fuel Handbook. West Conshohocken, PA: Infinity Publishing.

Ecker, Martin D. (1981) Radiation: All You Need to Know About It to Stop Worrying—Or to Start. New York, NY: Vintage Books.

The Economist. (2006) "The Heat is On: The Uncertainty Surrounding Climate Change Argues for Action, not Inaction—America Should Lead the Way." Editorial. September 7. Available at <http://www.ecoglobe.ch/climate/e/econ6909.htm>. Accessed March 3, 2008.

The Economist. (2005) "Cars in China." June 2. Available at <http://www.economist.com/world/displaystory.cfm?story_id=E1_QDPNRQN>. Accessed on March 13, 2008.

Eilperin, Juliet. (2006) "Growing Acidity of Oceans May Kill Corals." Washington Post, July 5, page A1. Available at <http://www.washingtonpost.com/wp-dyn/content/article/2006/07/04/AR2006070400772.html>. Accessed March 14, 2008.

Ezzati, M., A. Lopez, A. Rodgers, S. Van Der Hoorn, and C. Murray. (2002) "Selected Major Risk Factors and Global and Regional Burden of Disease." Lancet, volume 360, issue 9343: 1347–1360. Available at <http://www.thelancet.com/journals/lancet/article/PIIS0140673602114036/abstract>. Accessed March 14, 2008.

Ezzati, M., A. Lopez, A. Rodgers, and C. Murray. (2004) Comparative Quantification of Health Risks: Global and Regional Burden of Disease Attributable to Selected Major Risk Factors. Geneva, Switzerland: World Health Organization.

Fischlowitz-Roberts, Bernie. (2002) "Air Pollution Fatalities Now Exceed Traffic Fatalities by 3 to 1." Earth Policy Institute Eco-Economy Updates, number 13, September 17. Available at <www.earth-policy.org/Updates/Update17.htm>. Accessed February 28, 2008.

Glick, Daniel. (2004) "The Big Thaw: The Climate is Changing at an Unnerving Pace. Glaciers are Retreating, Ice Shelves are Fracturing, Sea Level is Rising, Permafrost is Melting. What Role Do Humans Play?" National Geographic, September. Available at <http://findarticles.com/p/articles/mi_hb3343/is_200409/ai_n12951395>. Accessed March 3, 2008.

Gordon, Greg. (2006) "Plug-in Hybrid Vehicles' Lobby Introduces Itself: A Coalition of Interests Wants to Show

There's a Demand for the Highly Efficient Autos to Reduce U.S. Oil Consumption." *Minneapolis StarTribune*, January 25. Available at <http://www.calcars.org/ calcars-news/262.html>. Accessed March 13, 2008.

Gore, Al. (2006) *An Inconvenient Truth: The Planetary Emergency of Global Warming and What We Can Do About It*. New York, NY: Rodale Books.

Gregory, Jonathan M., Philippe Huybrechts, and Sarah C.B. Raper. (2004) "Climatology: Threatened Loss of the Greenland Ice-sheet." *Nature*, volume 428, number 6983, April 8: 616. Available at <http://www.nature.com/nature/journal/ v428/n6983/abs/428616a.html;jsessionid =C6DA9FD9A3617AAE9F35FFDBF843 B586>. Accessed March 5, 2008.

Harvey, Fiona, and Stephen Fidler. (2007) "Industry Caught in Carbon 'Smokescreen'." *Financial Times*, April 25. Available at <http://www.ft.com/ cms/s/0/48e334ce-f355-11db-9845-000b5df10621.html?nclick_check=1>. Accessed March 9, 2007.

Hayden, Howard C. (2004) *The Solar Fraud: Why Solar Energy Won't Run the World*. Pueblo, CO: Vales Lake Publishing.

Hirschberg, Stefan, Peter Burgherr, and Alistair Hunt. (1998) "Accident Risks in the Energy Sector: Comparison of Damage Indicators and External Costs." Wuerenlingen and Villigen, Switzerland: Paul Scherrer Institute. Available at <http://gabe.web.psi.ch/pdfs/ PSAM7/0751.pdf>. Accessed March 14, 2008.

International Climate Change Taskforce. (2005) *Meeting the Climate Challenge: Recommendations of the International Climate Change Taskforce*. Washington, D.C.: Center for American Progress. [Simultaneously published by Institute for Public Policy Research (London, U.K.) and The Australia Institute (Canberra, Australia)]. Available at <http://www. americanprogress.org/kf/climatechallenge. pdf>. Accessed March 14, 2008.

Jacoby, Mitch. (2005) "Competing Visions of a Hydrogen Economy: Romm and Chalk Take Sides Over Research Priorities and Hydrogen's Role in Future Transportation and Energy Security Issues." *Chemical and Engineering News*, volume 83, number 34, August 22: 30–35. Available at <http://pubs.acs.org/cen/ government/83/8334hydrogen.html>. Accessed March 13, 2008.

Jha, Alok. (2006) "Planting Trees to Save Planet is Pointless, Say Ecologists." *Guardian* [U.K.], December 15. Available at <http://environment.guardian.co.uk/ print/0,,329663302-121568,00.html>. Accessed March 6, 2008.

Kahn, Joseph, and Mark Landler. (2007) "China Grabs West's Smoke-Spewing Factories." *New York Times*, December 21, A1. Available at <http://www.nytimes. com/2007/12/21/world/asia/21transfer. html>. Accessed March 14, 2008.

Kiley, David. (2007) "Big Oil's Big Stall on Ethanol." *Business Week*, September 23. Available at <http://www.businessweek. com/magazine/content/07_40/b4052052. htm?chan=search>. Accessed March 11, 2008.

Kindall, Henery, and David Pimentel. (1994) "Constraints on the Expansion of the Global Food Supply." *Ambio: The Official Publication of the Royal Swedish Academy of Sciences*, volume 23, number 3, May. Available at <http://dieoff.org/page36. htm>. Accessed March 12, 2008.

Kluger, Jeffrey. (2006) "Global Warming Heats Up." *Time*, volume 167, number 14, April 3. Available at <http:// www.time.com/time/magazine/ article/0,9171,1176980,00.html>. Accessed March 14, 2008.

Krzyzanowski, Michal. (2007) "Exposure of Children to Air Pollution (Particulate Matter) in Outdoor Air." Factsheet number 3.3. May. Code: RPG3_Air_Ex2_ PM. Bonn, Germany: WHO European Center for Environment and Health. Available at <http://www.euro.who.int/ Document/EHI/ENHIS_Factsheet_3_3. pdf>. Accessed March 4, 2008.

Künzli, N., R. Kaiser, S. Medina, M. Studnicka, O. Chanel, P. Filliger, M. Herry, F. Jorak Jr., V. Puybonnieux-Texier, P. Quenel, J. Schneider, R. Seethaler, J-c Vergnaud, and H. Sommer. (2000) "Public Health Impact of Outdoor and Traffic-related Air Pollution: A European Assessment." *Lancet*, 356: 795-801. Available at <http://www.

thelancet.com/journals/lancet/article/PIIS0140673600026532/abstract>. Accessed March 14, 2008.

Larsen, Janet. (2004) "Coal Takes Heavy Human Toll: Some 25,100 U.S. Deaths from Coal Use Largely Preventable." Earth Policy Institute, *Eco-economy Update*, August 24. Available at <www.earth-policy.org/Updates/Update42.htm>. Accessed February 27, 2008.

Lee, Jennifer. (2004) "EPA Raises Estimate of Babies Affected by Mercury Exposure." *New York Times*, February 10. Available at <http://query.nytimes.com/gst/fullpage.html?res=9B07E3DA173AF933A25751C0A9629C8B63>. Accessed March 4, 2008.

Leeb, Stephen, and Donna Leeb. (2005) *The Oil Factor: Protect Yourself and Profit from the Coming Energy Crisis*. New York, NY: Warner Business Books [renamed Business Plus in 2007].

Loof, Susanna. (2005) "Chernobyl Toll May Be Less than Feared." Associated Press, September 5. Available at <http://play.tm/wire/click/462567>. Accessed March 12, 2008.

Loucks, Robert A. (2002) *Shale Oil: Tapping the Treasure*. Philadelphia, PA: Xlibris Corporation.

Lovejoy, Thomas E. (2006) *Climate Change and Biodiversity*. New Haven, CT: Yale University Press.

Lovelock, James E. (2004) "Nuclear Power is the Only Green Solution." *The Independent* [London, U.K.], May 24. Available at <http://www.independent.co.uk/opinion/commentators/james-lovelock-nuclear-power-is-the-only-green-solution-564446.html>. Accessed March 11, 2008.

Markey, Sean. (2006) "Cold-Water Corals at Risk from Fishing Nets, Acidic Oceans." *National Geographic News*, May 1. Available at <http://news.nationalgeographic.com/news/2006/05/0501_060501_coral.html>. Accessed March 4, 2008.

Marsh, Gerald. (2002) "A Global Warming Primer." *National Policy Analysis*, number 420, July: 1–18. [An update of National Policy Analysis number 361, September 2001.] Available at <http://www.nationalcenter.org/NPA420.pdf>. Accessed March 4, 2008.

Mater, Terra (aka Matt Scherr). (2007) "Wasting My Money on Wind?" *Vail Daily* [Vail, Colorado], Feb 16. Available at <http://www.vaildaily.com/article/20070216/COLUMS/102130057&SearchID=73311024879897>. Accessed March 9, 2008.

Maurellis, Ahilleas. (2001) "Could Water Vapour Be the Culprit in Global Warming?" *Physics World*, volume 14, February 1: 22–23. Available at <http://physicsworld.com/cws/article/print/291>. Accessed March 14, 2008.

*Megawatt Motorworks*. (2006) "Subaru Canada Unveils Fast-Charging R1e Electric Prototype." *Megawatt Motorworks*, February 16. Available at <http://www.megawattmotorworks.com/display.asp?dismode=article&artid=246>. Accessed March 13, 2008.

Meneley, Dan. (2006) "Transition to Large Scale Nuclear Energy Supply." Presentation at twenty-seventh annual meeting, Canadian Nuclear Society. Available at <http://www.inea.org.br/Meneleypaper.htm>. Accessed March 13, 2008.

Michaels, Patrick J. (2005) *Meltdown: The Predictable Distortion of Global Warming by Scientists, Politicians, and the Media*. Washington, D.C.: The Cato Institute.

Ministry of Foreign Affairs [Japan]. Statement of support for Global Nuclear Energy Partnership. Available at <http://www.sustainablenuclear.org/PADs/pad0606reis.pdf>. Accessed March 12, 2008.

Monckton, Christopher. (2006a) "The Sun is Warmer Now than for the Past 11,400 Years." [Originally entitled "Climate Chaos? Don't Believe It."] *Sunday Telegraph* [London, U.K.], November 5. Available at <http://www.telegraph.co.uk/news/main.jhtml?xml=/news/2006/11/05/nwarm05.xml>. Accessed March 5, 2008.

Monckton, Christopher. (2006b) "Apocalypse Cancelled: Discussion, Calculations, and References." [This is a lengthy set of calculations to supplement Monckton's newspaper article of November 5, 2006.] Available at <http://www.telegraph.co.uk/news/graphics/2006/11/05/warm-refs.pdf>. Accessed March 5, 2008.

Moore, Patrick. (2005) "Statement to the Subcommittee on Energy and Resources, Committee on Government Reform, U.S. House of Representatives." Congressional testimony. April 28. Available at <http://www.nei.org/newsandevents/speechesandtestimony/2005/housemoore/>. Accessed March 12, 2008.

Morell, Virginia. (2004) "Now What? What Do You Get When You Compare Hundreds of Thousands of Years of Climate Data From Glaciers, Caves, and Coral Reefs with Climate Projections Modeled by the World's Most Powerful Supercomputers?" *National Geographic* (September). Available at <http://findarticles.com/p/articles/mi_hb3343/is_200409/ai_n12951393>. Accessed March 3, 2008.

Morris, Robert C. (2000) *The Environmental Case for Nuclear Power: Economic, Medical, and Political Considerations.* St. Paul, MN: Paragon House.

Murray, C., and A. Lopez. (1996) *The Global Burden of Disease: A Comprehensive Assessment of Mortality and Disability from Disease, Injuries, and Risk Factors in 1990 and Projected to 2020.* Cambridge, MA: Harvard University Press.

National Hydropower Association. (2008) National Hydropower Association homepage. Available at <http://www.hydro.org/>. Accessed March 13, 2008.

National Petroleum Council. (1999) *Meeting the Challenges of the Nation's Growing Natural Gas Demand,* vols. 1–3. Washington, D.C.: National Petroleum Council.

*New York Times.* (2006) "India, Oil and Nuclear Weapons." Editorial. February 19. Available at <http://www.nytimes.com/2006/02/19/opinion/19sun1.html>. Accessed March 10, 2008.

Norris, Scott. (2007) "Acid Oceans Threatening Marine Food Chain, Experts Warn." *National Geographic News,* February 17. Available at <http://news.nationalgeographic.com/news/2007/02/070217-acid-oceans.html>. Accessed March 4, 2008.

Ostro, Bart. (2004) *Outdoor Air Pollution: Assessing the Environmental Burden of Disease at National and Local Levels.* WHO Environmental Burden of Disease Series, Number 5. Geneva,

Switzerland: World Health Organization. Available at <http://whqlibdoc.who.int/publications/2004/9241591463.pdf>. Accessed March 14, 2008.

Pegg, J.R. (2004) "Coal Power Soot Kills 24,000 Americans Annually." *Environment News Service,* June 10. Available at <http://www.ens-newswire.com/ens/jun2004/2004-06-10-10.asp>. Accessed February 29, 2008.

Pickrell, John. (2004) "Oceans Found to Absorb Half of All Man-Made Carbon Dioxide." *National Geographic News,* July 15. Available at <http://news.nationalgeographic.com/news/2004/07/0715_040715_oceancarbon.html>. Accessed March 4, 2008.

Pimentel, David, and Tad W. Patzek. (2005) "Ethanol Production Using Corn, Switchgrass, and Wood; Biodiesel Production Using Soybean and Sunflower." *Natural Resources Research,* volume 14, number 1, March: 65–76. Available at <http://www.springerlink.com/content/r1552355771656v0/>. Accessed March 11, 2008.

Population Institute. (no date, likely 2007) Issues Modules page. Available at <http://www.populationinstitute.org/cms/modules/PopIssues/front/lib/pdf.php?id=14>. Accessed March 4, 2008.

Population Institute. (2006) "Population Challenges: The Basics." Issue Brief, February 2006 update. Available at <http://www.populationinstitute.org/cms/modules/PopIssues/front/lib/pdf.php?id=5>. Accessed March 3, 2008.

Postel, Sandra. (1999) *Pillar of Sand: Can the Irrigation Miracle Last?* New York, NY: W.W. Norton and Company.

Princeton University, Office of Environmental Health and Safety. (2007) Open Source Radiation Safety Training, Module 2: "Background Radiation and Other Sources of Exposure." Available at <http://web.princeton.edu/sites/ehs/osradtraining/backgroundradiation/background.htm>. Accessed March 12, 2008.

Radiation Effects Research Foundation. (n.d., likely 2006) *US-Japan Joint Reassessment of Atomic Bomb Radiation Dosimetry in Hiroshima and Nagasaki: Final Report.* DS86. Hiroshima, Japan: Radiation

Effects Research Foundation. Available at <http://www.rerf.jp/shared/ds86/ds86a.html>. Accessed March 12, 2008.

Radiation Effects Research Foundation. (2005) *Reassessment of the Atomic Bomb Radiation Dosimetry in Hiroshima and Nagasaki: Dosimetry System 2002.* Hiroshima, Japan: Radiation Effects Research Foundation. Available at <http://www.rerf.jp/shared/ds02/index.html>. Accessed March 12, 2008.

Raloff, Janet. (2003) "Global Food Trends." *Science News Online*, volume 163, number 22, May 31. Available at <http://www.sciencenews.org/articles/20030531/food.asp>. Accessed March 12, 2008.

Ridenour, Amy. (2005) "Spinning Global Warming." *National Policy Analysis*, number 523, February. Available at <www.nationalcenter.org/NPA523GlobalWarmingPodesta.html>. Accessed April 15, 2008.

Sabine, Christopher et al. (2005a) "The Ocean Sink for Anthropogenic $CO_2$." *Science*, volume 308, June 17: 1740. Available at: <http://www.sciencemag.org/cgi/search?src=hw&site_area=sci&fulltext=%22christopher+sabine%22>. Accessed March 4, 2008.

Sabine, Christopher et al. (2005b) "Response to Comment on 'The Ocean Sink for Anthropogenic $CO_2$.'" *Science*, volume 308, June 17: 1743. Available at <http://www.sciencemag.org/cgi/search?src=hw&site_area=sci&fulltext=%22christopher+sabine%22>. Accessed March 4, 2008.

Smith, J.W. (1980) "Oil Shale Resources of the United States." *Mineral and Energy Resources*, volume 23, number 6.

Socolow, Robert H., and Stephen W. Pacala. (2006) "A Plan to Keep Carbon in Check." *Scientific American*, September: 50–57. Available at <http://www.sciam.com/carbon/0906050.pdf>. Accessed March 13, 2008.

Solar Energy Industries Association (SEIA). (2004) "Our Solar Power Future: The U.S. Photovoltaics Industry Roadmap Through 2030 and Beyond." Washington, D.C.: SEIA. Available at <www.seia.org/roadmap.pdf>. Accessed March 11, 2008.

Souers, Amy. (2000) "American Rivers: Damning the Dams." *eMagazine*, July 1.

Available at <http://www.emagazine.com/view/?309&printview&src=>. Accessed March 13, 2008.

Stark, L., ed. (2003) *Vital Signs 2003.* New York, NY: W.W. Norton. Available at <http://www.worldwatch.org/node/1056>. Accessed March 12, 2008.

Steyn, Mark. (2007) "The Eco-Messiah's Literally Glowing Example." *Orange County Register*, March 4. Available at <http://www.ocregister.com/ocregister/opinion/nationalcolumns/article_1596690.php>. Accessed March 9, 2008.

Stipp, David. (2001) "The Coming Hydrogen Economy: Fuel Cells Powered by Hydrogen Are About to Hit the Market." *Fortune*, November 12. Available at <http://money.cnn.com/magazines/fortune/fortune_archive/2001/11/12/313316/index.htm>. Accessed March 13, 2008.

Svitil, Kathy A. (2005) "Discover Dialogue: Meteorologist William Gray." *Discover*, volume 26, September 9. Available at <http://discovermagazine.com/2005/sep/discover-dialogue/>. Accessed March 5, 2008.

Swisher, Erica. (2005) "The Real Cost of Oil." *Ethanol Today*, August. Available at <http://www.drivingethanol.org/userdocs/Real_Cost_of_Oil_Aug_05.pdf>. Accessed March 11, 2008.

Till, Charles E. (2007) "Reminiscences of Reactor Development at Argonne National Laboratory." W.B. Lewis Lecture, Canadian Nuclear Society annual meeting, Saint John, New Brunswick, Canada, June 4. Available at <http://www.ecolo.org/documents/documents_in_english/IFR-integral-fast-reactor07.pdf>. Accessed March 11, 2008.

Tilman, David, Jason Hill, and Clarence Lehman. (2006) "Carbon-Negative Biofuels from Low-Input High-Diversity Grassland Biomass." *Science*, volume 314, number 5805, December 8: 1598–1600. Available at <http://www.sciencemag.org/cgi/content/abstract/314/5805/1598>. Accessed March 11, 2008.

Tolley, George S., and Donald W. Jones. (2004) "The Economic Future of Nuclear Power." A study conducted at the University of Chicago. Chicago, IL. August. Available at <http://www.anl.gov/Special_Reports/

NuclEconSumAug04.pdf>. Accessed
March 14, 2008.

Vachhani, Ashish. (2005) "India's Energy
Security Dilemma." *Hindu Business Line*,
April 26. Available at <http://www.
thehindubusinessline.com/2005/04/26/
stories/2005042600270800.htm>.
Accessed March 10, 2008.

Varchaver, Nicholas. (2005) "Nuclear Spring?:
Some Scientists Are Learning to Love
America's Most Reviled Source of Energy."
*Fortune*, January 10. Available at <http://
money.cnn.com/magazines/fortune/
fortune_archive/2005/01/10/8230954/
index.htm>. Accessed March 11, 2008.

Veziroglu, T. Nejat. (2000) Comments at
opening press conference of Hyforum
2000. Available at <http://www.
anticipation.info/texte/buckminster/
www.bfi.org/Trimtab/fall00/hyforum.
htm>. Accessed March 13, 2008.

*Wall Street Journal*. (2007) "Climate of
Opinion: The Latest U.N. Report Shows
the 'Warming' Debate is Far From
Settled." Editorial. February 5. Available
at: <http://opinionjournal.com/editorial/
feature.html?id=110009625>. Accessed
March 4, 2008.

Weisman, Alan. (2007) *The World Without Us*.
New York, NY: Thomas Dunne Books.

*Who Killed the Electric Car?* (2006) Directed by
Chris Paine. Produced by Dean Devlin.
Columbia TriStar Films/Sony Pictures
Classics.

Winsor, P. (2001) "Arctic Sea Ice Thickness
Remained Constant During the 1990s."
*Geophysical Research Letters*, volume 28,
number 6, March 15: 1039–1041. Available
at <http://www.whoi.edu/science/PO/
people/pwinsor/pdfs/winsor_2001.pdf>.
Accessed March 5, 2008.

World Energy Council. (2007a) *Survey of
Energy Sources 2007*. New York, NY:
Palgrave Macmillan. Available at <http://
www.worldenergy.org/publications/
survey_of_energy_resources_2007/
default.asp>. Accessed March 9, 2008.

World Energy Council. (2007b) *Energy Policy
Scenarios to 2050*. New York, NY: Palgrave
Macmillan. Available at <http://www.
worldenergy.org/publications/energy_
policy_scenarios_to_2050/default.asp>.
Accessed March 9, 2008.

Worldwatch Institute. (1999) "Emerging
Water Shortages Threaten Food Supplies,
Regional Peace." July 17. Washington,
D.C.: Worldwatch Institute. Available
at <http://www.worldwatch.org/
node/1654>. Accessed March 12, 2008.

Zimov, Sergey A., Edward A. G. Schuur,
and F. Stuart Chapin III. (2006)
"Climate Change: Permafrost and the
Global Carbon Budget." *Science*, volume
312, number 5780, June 16: 1612–1613.
Available at <http://www.sciencemag.org/
cgi/content/summary/312/5780/1612>.
Accessed March 5, 2008.

Zuzel, Michael. (2000) "Dam Nation:
Reservoirs of Controversy."
*IntellectualCapital.com*, June 8. Available
<http://www.bluefish.org/damnatin.
htm>. Accessed March 13, 2008.

## United Nations Sources

United Nations Environment Programme.
(2003) "Power Stations Threaten People
and Wildlife with Mercury Poisoning."
UNEP press release. February 3. Available
at <http://www.unep.org/Documents.
Multilingual/Default.asp?DocumentID=
284&ArticleID=3204&l=en>. Accessed
February 27, 2008. Full report available
at <www.unep.org/GoverningBodies/
GC22>.

United Nations Intergovernmental Panel on
Climate Change. (2007) *Climate Change
2007: Synthesis Report*. Fourth Assessment
Report. New York, NY: United Nations.
Available at <http://www.ipcc.ch/> and at
<http://www.ipcc.ch/ipccreports/ar4-syr.
htm>. Both accessed March 4, 2008.

United Nations International Atomic
Energy Agency, International Nuclear
Safety Advisory Group [INSAG]. (1992)
"INSAG-7: The Chernobyl Accident—
Updating of INSAG-1, A report by the
International Safety Advisory Group."
*Safety Series*, number 75-INSAG-7. Vienna,
Austria: International Atomic Energy
Agency. Available at <http://www-pub.
iaea.org/MTCD/publications/PDF/
Pub913e_web.pdf>. Accessed March 12,
2008.

United Nations International Atomic Energy
Agency. (2005) "Chernobyl: The True
Scale of the Accident." New York, NY:
United Nations. Available at <http://

www.iaea.org/NewsCenter/Focus/Chernobyl/pdfs/pr.pdf>. Accessed March 12, 2008.

United Nations Scientific Commission on the Effects of Atomic Radiation [UNSCEAR]. (2000) *Sources and Effects of Ionizing Radiation: UNSCEAR 2000 Report to the General Assembly, with Scientific Annexes, Volume 2: Effects,* Annex J: "Exposures and Effects of the Chernobyl Accident." Available at <http://www.uic.com.au/unscearcherno.htm>. Accessed March 12, 2008.

United Nations World Health Organization. (2007) "Chronic Obstructive Pulmonary Disease (COPD)." Factsheet number 315, November. Available at <http://www.who.int/mediacentre/factsheets/fs315/en/index.html>. Accessed February 28, 2008.

United Nations World Health Organization. (2005a) "Indoor Air Pollution and Health." Factsheet number 292. Available at <http://www.who.int/mediacentre/factsheets/fs292/en/>. Accessed March 4, 2008.

United Nations World Health Organization. (2005b) "Particulate Matter Air Pollution: How It Harms Health." Factsheet EURO/04/05. April 14. Available at <http://www.euro.who.int/document/mediacentre/fs0405e.pdf>. Accessed March 4, 2008.

United Nations World Health Organization. (2003) "Deaths and DALYs Attributable to Outdoor Air Pollution." Available at <http://www.who.int/quantifying_ehimpacts/national/countryprofile/mapoap/en/>. Accessed March 4, 2008.

United Nations World Water Assessment Programme. (2003) *First UN World Water Development Report: Water for People, Water for Life.* New York, NY: United Nationsl Educational, Scientific, and Cultural Organization [UNESCO] and Berghahn Books. Available at <http://www.unesco.org/water/wwap/wwdr1/>. Accessed March 13, 2008.

## U.S. Government Sources

Census Bureau. (2007) "Total Midyear Population for the World: 1950–2050." Available at <http://www.census.gov/ipc/www/idb/worldpop.html>. Accessed March 3, 2008.

Centers for Disease Control and Prevention. (2005) *CDC's Third National Report on Human Exposure to Environmental Chemicals: Spotlight on Mercury.* NCEH Publication 05–0664, July. Atlanta, GA: Centers for Disease Control and Prevention. Available at <http://www.cdc.gov/exposurereport/pdf/factsheet_mercury.pdf>. Accessed February 28, 2007.

Centers for Disease Control and Prevention. (2004) "Blood Mercury Levels in Young Children and Childbearing-Aged Women: United States, 1999–2002." *Morbidity and Mortality Weekly Report,* volume 53, number 43, November 5: 1018–1020. Atlanta, GA: Centers for Disease Control and Prevention.

Central Intelligence Agency. (2004) *The World Fact Book 2004.* Washington, D.C.: Government Printing Office. Available at <http://www.umsl.edu/services/govdocs/wofact2004/index.html> Accessed March 9, 2007.

Department of Energy, Office of Deputy Assistant Secretary for Petroleum Reserves, and Office of Naval Petroleum and Oil Shale Reserves. (2004) *Strategic Significance of America's Oil Shale Resource, volume II: Oil Shale Resources Technology and Economics.* Washington, D.C.: U.S. Department of Energy. Available at <http://www.fe.doe.gov/programs/reserves/npr/publications/npr_strategic_significancev2.pdf>. Accessed March 13, 2008.

Department of Energy, Office of Energy Efficiency and Renewable Energy, National Renewable Energy Laboratory. (2004) *Solar Energy Technologies Program: Multi-Year Technical Plan, 2003–2007 and Beyond.* Washington, D.C.: Government Printing Office. Available at <http://www.nrel.gov/docs/fy04osti/33875.pdf>. Accessed March 10, 2008.

Environmental Protection Agency. (2006) "Global Warming News and Events: Science and Policy News." Occasionally updated through March 24, 2006. Available at <http://yosemite.epa.gov/oar/globalwarming.nsf/content/NewsandEventsScienceandPolicyNews.html>. Accessed March 5, 2008. [Since March 2006, relevant information is added to the EPA's "Climate Change" website, available at <http://epa.gov/climatechange/index.html>.]

Environmental Protection Agency. (2005) "Controlling Power Plant Emissions: Overview." EPA press release concerning EPA's Clean Air Mercury Rule. Available at <www.epa.gov/hg/control_emissions/index.htm> (accessed February 27, 2008). Fuller information on the Clean Air Mercury Rule available at <www.epa.gov/air/mercuryrule/basic.htm>.

Environmental Protection Agency. (1997) *Mercury Study Report to Congress*, volumes I-VIII. Washington, D.C.: Environmental Protection Agency. Available at <http://www.epa.gov/hg/report.htm>. Accessed March 14, 2008.

Environmental Protection Agency. (n.d., but continuously updated) "Human Exposure [to Mercury]." Available at <http://www.epa.gov/hg/exposure.htm>. Last updated February 25, 2008. Accessed February 27, 2008.

Lawrence Berkeley National Laboratory. (n.d., likely February 1999) "Natural Sources of Radioactivity." Available at <http://ap.lbl.gov/LBL-Programs/tritium/natural-dosage.html>. Accessed March 12, 2008.

National Research Council, Advisory Committee on the Biological Effects of Ionizing Radiations. (1980) *Effects on Populations of Exposure to Low Levels of Ionizing Radiation*. Washington, D.C.: National Academies Press.

National Research Council, Committee on Environmental Impacts of Wind Energy Projects. (2007) *Environmental Impacts of Wind-energy Projects*. Washington, D.C.: National Academies Press. Available at <http://books.nap.edu/openbook.php?isbn=0309108349&page=R1>. Accessed March 11, 2008.

National Research Council, Committee on the Toxicological Effects of Methylmercury. (2000) *Toxicological Effects of Methylmercury*. Washington, D.C.: National Academies Press.

National Snow and Ice Data Center. (n.d.) "All About Glaciers: Quick Facts." Available at <http://nsidc.org/glaciers/quickfacts.html>. Accessed March 5, 2008.

Oak Ridge National Laboratory, in collaboration with U.S. Department of Agriculture, USDA Agricultural Research Service, USDA Forest Service, USDA Office of the Chief Economist, Office of Energy Policy and New Uses, and U.S. Department of Energy. (2005) *Biomass as Feedstock for a Bioenergy and Bioproducts Industry: The Technical Feasibility of a Billion-Ton Annual Supply*. A joint study sponsored by the U.S. Department of Energy, Office of Energy Efficiency and Renewable Energy, and the Office of the Biomass Program. Prepared for the U.S. Department of Agriculture and the U.S. Department of Energy. April. Available at <http://feedstockreview.ornl.gov/pdf/billion_ton_vision.pdf>. Accessed March 14, 2008.

United States Geological Survey. (2007) "Water Use in the United States." *National Atlas of the United States*. Washington, D.C.: U.S. Department of the Interior. Available at <http://nationalatlas.gov/articles/water/a_wateruse.html>. Accessed March 13, 2008.

United States Geological Survey. (2000) "Mercury in the Environment." Factsheet 146–00 (October). Available at <http://www.usgs.gov/themes/factsheet/146–00/index.html>. Accessed February 28, 2008.

## Other Valuable Sources

Abt Associates. (2000) *Death, Disease, and Dirty Power: Mortality and Health Damage Due to Air Pollution from Power Plants*. A report prepared for the Clean Air Task Force. Washington, D.C. and Cambridge, MA: Abt Associates. Available at <http://static.uspirg.org/reports/dirtypower2000.pdf>. Accessed March 4, 2008.

Anderson, Terry L., and Donald Leal. (1997) *Enviro-Capitalists*. New York, NY: Rowman and Littlefield.

Bagwell, Kyle, and Robert W. Staiger. (2004) *The Economics of the World Trading System*. New edition. Cambridge, MA: The MIT Press.

Bailey, Ronald. (1993) *Eco-scam: The False Prophets of Ecological Apocalypse*. New York, NY: St. Martin's Press.

Barber, Benjamin. (2004) *Fear's Empire: War, Terrorism, and Democracy*. New York, NY: W.W. Norton and Company.

Bast, Joseph L., and James M. Taylor. (2007) *Scientific Consensus on Global Warming: Results of an International Survey of Climate Scientists*. Second edition, revised. Chicago, IL: The Heartland Institute.

Bedard, Patrick. (2006) "2007 Saturn Vue Green Line: Short Take Road Test." *Car and Driver*, November. Available at <http://www.caranddriver.com/reviews/hot_lists/high_performance/performance_files_tested_by_c_d/2007_saturn_vue_green_line_short_take_road_test>. Accessed March 13, 2008.

Bedard, Patrick. (2005) "Buying Pleasure: Will it be a Hemi or a Hybrid?" *Car and Driver*, December. Available at <http://www.calcars.org/calcars-news/199.html>. Accessed March 13, 2008.

Bergeron, Louis. (2008) "Study Links Carbon Dioxide Emissions to Increased Deaths." *Stanford Report*, January 3. Available at <http://news-service.stanford.edu/news/2008/january9/co-010908.html>. Accessed February 28, 2008.

Blanchard, Roger D. (2005) *The Future of Global Oil Production: Facts, Figures, Trends, and Projections by Region*. Jefferson, NC: McFarland and Company.

Brown, Lester. (2006) *Plan B 2.0: Rescuing a Planet Under Stress and a Civilization in Trouble*. Expanded update edition. New York, NY: W.W. Norton and Company.

Campbell, C.J. (2005) *Oil Crisis*. Brentwood, UK: Multi-Science Publishing, Ltd.

Campbell, C.J. (1988/2004) *The Coming Oil Crisis*. Essex, U.K.: Multi-Science Publishing Company and Petroconsultants S.A.

Carbon, Max. (1997) *Nuclear Power: Villain or Victim?: Our Most Misunderstood Source of Electricity* Madison, WI: Pebble Beach Publishers.

Carmichael, Mary. (2007) "Troubled Waters." *Newsweek*, June 4. Available at <http://www.globalpolicy.org/socecon/gpg/2007/0604troubled.htm>. Accessed March 13, 2008.

Clean Air Task Force. (2005) "Diesel and Health in America: The Lingering Threat." Boston, MA: Clean Air Task Force. Available at <http://www.catf.us/publications/view/83>. Accessed February 29, 2008.

Cole, Nancy, and P.J. Skerrett. (1995) *Renewables are Ready: People Creating Renewable Energy Solutions*. White River Junction, VT: Chelsea Green Publishing Company.

Consumeraffairs.com. (2005) "EPA Fuel Economy Ratings Have Shortfalls of Up to 50 Percent: Window Stickers Vastly Overstate MPG, *Consumer Reports* Study Finds." Consumeraffairs.com, September 14. Available at <http://www.consumeraffairs.com/news04/2005/cu_gas_mileage.html>. Accessed March 13. 2008.

Costanza, Robert. (1991) *Ecological Economics: The Science and Management of Sustainability*. New York, NY: Columbia University Press.

Costanza, Robert, John H. Cumberland, Herman Daly, Robert Goodland, and Richard B. Norgaard. (1997) *An Introduction to Ecological Economics*. Boca Raton, FL: CRC Press.

Darley, Julian. (2004) *High Noon for Natural Gas: The New Energy Crisis*. White River Junction, VT: Chelsea Green Publishing Company.

Davies, J. Clarence III. (1974) *Energy Politics*. New York, NY: St. Martin's Press.

Deffeyes, Kenneth S. (2005) *Beyond Oil: The View from Hubbert's Peak*. New York, NY: Hill and Wang.

Deffeyes, Kenneth S. (2003) *Hubbert's Peak: The Impending World Oil Shortage*. Princeton, NJ: Princeton University Press.

Eilperin, Juliet. (2006) "World's Fish Supply Running Out, Researchers Warn." *Washington Post*, November 3, page A1. Available at <http://www.washingtonpost.com/wp-dyn/content/article/2006/11/02/AR2006110200913.html>. Accessed March 14, 2008.

Friedman, Thomas. (2005) *The World is Flat: A Brief History of the Twenty-first Century*. New York, NY: Farrar, Straus, and Giroux.

Garrett, Major and Timothy Penny. (1998) *The 15 Biggest Lies in Politics*. New York, NY: St. Martin's Griffin.

Geller, Howard. (2003) *Energy Revolution: Politics for a Sustainable Future*. Washington, D.C.: Island Press.

Gelbspan, Ross. (1997) *The Heat is On: The Climate Crisis, the Cover-Up, the Prescription*. New York, NY: Perseus Books.

Gibbard, S., K. Caldeira, G. Bala, T.J. Phillips, and M. Wickett. (2005) "Climate Effects of Global Land Cover Change." *Geophysical*

*Research Letters*, vol. 32, L23705, December 8. Available at <http://www.agu.org/pubs/crossref/2005/2005GL024550.shtml>. Accessed March 6, 2008.

Goodell, Jeff. (2006) *Big Coal: The Dirty Secret Behind America's Energy Future*. New York, NY: Houghton Mifflin.

Goodstein, David. (2005) *Out of Gas: The End of the Age of Oil*. New York, NY: W. W. Norton.

Grant, Lindsey. (2005) *The Collapsing Bubble: Growth and Fossil Energy*. Santa Ana, CA: Seven Locks Press.

Hatch, M., E. Ron, A. Bouville, L. Zablotska, and G. Howe. (2005) "The Chernobyl Disaster: Cancer Following the Accident at the Chernobyl Nuclear Power Plant." *Epidemiological Review*, volume 27: 56–66.

Heaberlin, Scott W. (2004) *A Case for Nuclear-Generated Electricity*. Richland, WA: Battelle Press. [A publication of the Pacific Northwest National Laboratory.]

Heinberg, Richard. (2004) *Power Down: Options and Actions for a Post-Carbon World*. Gabriola Island, British Columbia, Canada: New Society Publishers.

Heinberg, Richard. (2003) *The Party's Over: Oil, War, and the Fate of Industrial Societies*. Revised second edition in 2005. Gabriola Island, British Columbia, Canada: New Society Publishers.

Hoff, Mary. (2008) "Land of Biofuels?: Can We Get Gas from Grasses?" *Minnesota Conservation Volunteer*, volume 71, number 416, January-February: 9–21. Available at <http://www.dnr.state.mn.us/volunteer/janfeb08/biofuels.html>. Accessed March 14, 2008.

Howard, Philip. (1994) *The Death of Common Sense: How Law is Suffocating America*. Clayton, Victoria, Australia: Warner Books Pty Ltd.

Howe, G.R., L.B. Zablotska, J.J. Fix, J. Egel, and J. Buchanan. (2004) "Analysis of the Mortality Experience Amongst U.S. Nuclear Power Industry Workers After Chronic Low-Dose Exposure to Ionizing Radiation." *Radiation Research*, volume 162, issue 5, November: 517-526. Available at <http://www.bioone.org/perlserv/?request=getabstract&doi=10.1667%2FRR3258&ct=1>. Accessed March 12, 2008.

Huber, Peter. (1999) *Hard Green: Saving the Environment from the Environmentalists—A Conservative Manifesto*. New York, NY: Basic Books.

Huber, Peter W. and Mark P. Mills. (2005) *The Bottomless Well: The Twilight of Fuel, the Virtue of Waste, and Why We Will Never Run Out of Energy*. New York, NY: Basic Books.

Jacobson, Mark Z. (2008) "On the Causal Link Between Carbon Dioxide and Air Pollution Mortality." *Geophysical Research Letters*, vol. 35, L03809, February 12. available at <http://www.stanford.edu/group/efmh/jacobson/2007GL031101.pdf>. Accessed February 28, 2008.

Kaku, Michio and Jennifer Trainer. (1982) *Nuclear Power: Both Sides*. New York, NY: W. W. Norton.

Klare, Michael T. (2004) *Blood and Oil*. New York, NY: Owl Books.

Klare, Michael T. (2001) *Resource Wars*. New York, NY: Henry Holt and Company.

Krabill, W. et al. (2004) "Greenland Ice Sheet: Increased Coastal Thinning." *Geophysical Research Letters*, vol. 31, L24402. Available at <http://www.agu.org/pubs/crossref/2004/2004GL021533.shtml>. Accessed March 5, 2008.

Kunstler, James Howard. (2005) *The Long Emergency: Surviving the End of Oil, Climate Change, and Other Converging Catastrophes of the Twenty-first Century*. New York, NY: Atlantic Monthly Press.

Leahy, Elizabeth, with Robert Engelman, Carolyn Gibb Vogel, Sarah Haddock, and Tod Preston. (2007) *The Shape of Things to Come: Why Age Structure Matters to A Safer, More Equitable World*. Washington, D.C.: Population Action International. Available at <http://www.populationaction.org/Publications/Reports/The_Shape_of_Things_to_Come/Summary.shtml>. Accessed March 3, 2008.

Leahy, Stephen. (2006) "Global Food Supply Near the Breaking Point." *IPS-InterPress Service*, May 17. Available at <http://ipsnews.net/news.asp?idnews=33268>. Accessed March 12, 2008.

Leeb, Stephen and Glen Strathy. (2006) *The Coming Economic Collapse: How You Can Thrive When Oil Costs $200 a Barrel*. Warner Basic Books [renamed Business Plus in 2007].

Leggett, Jeremy. (2005) *The Empty Tank: Oil,*

*Gas, Hot Air, and the Coming Global Financial Catastrophe.* New York, NY: Random House.

Lewis, Charles. (2004) *The Buying of the President 2004: Who's Really Bankrolling Bush and His Democratic Challengers—and What they Expect in Return.* New York, NY: Harper Collins Perennial.

Lomborg, Bjorn. (2007) *Cool It: The Skeptical Environmentalist's Guide to Global Warming.* New York, NY: Knopf.

Lovelock, James E. (2007) "Go Nuclear, Save the Planet." *London Times,* Sunday, February 18. Available at <http://www.timesonline.co.uk/tol/news/uk/article1400073.ece>. Accessed March 11, 2008.

Mast, Tom. (2005) *Over a Barrel: A Simple Guide to the Oil Shortage.* Plymouth, MI: Hayden Publishers.

McKillop, Andrew, ed. (2005) *The Final Energy Crisis.* London, U.K.: Pluto Press.

Moore, Patrick. (2006) "Going Nuclear: A Green Makes the Case." *Washington Post,* April 16: page B1. Available at <http://www.washingtonpost.com/wp-dyn/content/article/2006/04/14/AR2006041401209.html>. Accessed March 12, 2008.

Moore, Patrick. (1994) "Hard Choices for the Environmental Movement." *Leadership Quarterly,* volume 5, numbers 3/4. Available at <http://www.greenspirit.com/key_issues/the_log.cfm?booknum=12>. Accessed March 12, 2008.

*National Geographic.* (2008) Special Report: "Changing Climate: What You Should Know, What You Can Do." June. Available at <www.climate.ngm.com>. Accessed March 27, 2007.

Ontario Medical Association. (2000) *The Illness Costs of Air Pollution.* Toronto, Canada: Ontario Medical Association. Available at <www.oma.org/phealth/icap.htm>. Accessed March 12, 2008.

Pearce, Fred. (2006) *When the Rivers Run Dry: Water—The Defining Crisis of the Twenty-first Century.* Boston, MA: Beacon Press.

Penny, Tim and Major Garrett. (1995) *Common Cents: A Retiring Six-Term Congressman Reveals How Congress Really Works and What We Must Do to Fix It.* New York, NY: Harper Collins Avon.

Peters, Charles. (1994) *How Washington Really Works.* Lexington, MA: Addison Wesley.

Pfeiffer, Dale Allen. (2006) *Eating Fossil Fuels: Oil, Food, and the Coming Crisis in Agriculture.* Gabriola Island, British Columbia, Canada: New Society Publishers.

*Plenty: The World in Green* [magazine]. (2008) "Sunny Forecast: Solar Energy on the Rise." April/May: 38. [No author listed.]

Posner, Richard. (2004) *Catastrophe: Risk and Response.* New York, NY: Oxford University Press USA.

Ray, Dixy Lee with Lou Guzzo. (1993) *Environmental Overkill: Whatever Happened to Common Sense?* New York, NY: Harper Collins Perennial.

Rees, Sir Martin. (2003) *Our Final Hour: A Scientist's Warning—How Terror, Error, and Environmental Disaster Threaten Humankind's Future in This Century ... On Earth and Beyond.* New York, NY: Basic Books.

Roberts, Paul. (2004) *End of Oil: On the Edge of a Perilous New World.* Boston, MA: Mariner Books.

Rosegrant, Mark W., Ximing Cai, and Sarah A. Cline. (2002) *World Water and Food to 2025: Dealing with Scarcity.* Washington, D.C.: International Food Policy Research Institute. Available at <http://www.ifpri.org/pubs/books/water2025/water2025.pdf>. Accessed March 12, 2008.

Rothman, Stanley, and S. Robert Lichter. (1987) "Elite Ideology and Risk Perception in Nuclear Energy Policy." *American Political Science Review,* volume 81, number 2, June. Available at <http://links.jstor.org/sici?sici=00030554(198706)81:2%3C383:EIARPI%3E2.0.CO;2-X>. Accessed March 12, 2008.

Ruppert, Michael C. (2004) *Crossing the Rubicon: The Decline of the American Empire at the End of the Age of Oil.* Gabriola Island, British Columbia, Canada: New Society Publishers.

Schlesinger, Victoria. (2008) "Electrifying Breakthroughs: Better Battery Technology Will Make Eco Cars Road-Ready." *Plenty: The World in Green,* April/May: 39.

Schurr, Sam. (1973) *Energy, Economic Growth, and the Environment.* Washington, D.C.: RFF Press.

Shabecoff, Philip. (1990) "Study Finds No Increased Risk of Cancer Deaths Near Nuclear Sites." *New York Times*, September 20. Available at <http://query.nytimes.com/gst/fullpage.html?res=9C0CE4D7173CF933A1575AC0A966958260&sec=&spon=&pagewanted=all>. Accessed March 12, 2008.

Shigematsu, Itsuzo. (1998) "Greetings: Fifty Years of Atomic Bomb Casualty Commisstion-Radiation Effects Research Foundation Studies." *Proceedings of the National Academy of Sciences of the United States of America*, volume 95, issue 10, May 12: 5424-5425. Available at <http://www.pnas.org/cgi/content/full/95/10/5424>. Accessed March 12, 2008.

Simmons, Matthew R. (2005) *Twilight in the Desert: The Coming Saudi Oil Shock and the World Economy*. New York, NY: John Wiley and Sons.

Stiglitz, Joseph E. (2005) *Fair Trade for All*. New York, NY: Oxford University Press USA.

Stroup, Richard L. (2003) *Eco-nomics*. Washington, D.C.: Cato Institute.

Swanenberg, August. (2005) *Macroeconomics Demystified*. New York, NY: McGraw Hill.

Tennesen, Michael. (2004) *The Complete Idiot's Guide to Global Warming*. New York, NY: Alpha Books.

Tertzakian, Peter. (2006) *A Thousand Barrels a Second: The Coming Oil Break Point and the Challenges Facing an Energy-Dependent World*. New York, NY: McGraw Hill.

Tickell, Joshua. (2003) *From the Fryer to the Fuel Tank*. New Orleans, LA: Joshua Tickell Publications.

United States Nuclear Regulatory Commission. (2004) "Factsheet on the Three Mile Island Accident." Available at <http://www.nrc.gov/reading-rm/doc-collections/fact-sheets/3mile-isle.html>. Accessed March 12, 2008.

United States Nuclear Regulatory Commission. (2000) "Factsheet on the Chernobyl Nuclear Power Plant." Available at <http://www.nrc.gov/reading-rm/doc-collections/fact-sheets/fschernobyl.html>. Accessed March 12, 2008.

Vaitheeswaran, Vijay V. (2008) "Go, Clean Racer, Go." *Plenty: The World in Green*, April/May: 76-83.

Weart, Spencer. (2003) *The Discovery of Global Warming*. Cambridge, MA: Harvard University Press.

Wells, Stewart. (2006) "An Open Letter to the United Nations." National Farmers Union [Canada], press release, May 9. Available at <http://www.nfu.ca/press_releases/press/2006/May_06/UN_FOOD_LETTER_FINAL.pdf>. Accessed March 12, 2008.

Whipple, Tom. (2006) "The Peak Oil Crisis: Ethanol and Peak Food." *Falls Church [VA] News-Press*, volume 16, number 12, May 25-31, online issue. Available at <http://www.fcnp.com/612/peakoil.htm>. Accessed March 12, 2008.

Wiggin, Addison. (2005) *The Great Shanghai Blackout of 2006*. Baltimore, MD: Agora Financial LLC.

Worldwatch Institute. (2006) *State of the World*. New York, NY: W. W. Norton.

Worldwatch Institute. (2005) *State of the World*. New York, NY: W. W. Norton.

Worldwatch Institute. (2005) *Vital Signs*. New York, NY: W. W. Norton.

Worldwatch Institute. (2004) *State of the World*. New York, NY: W. W. Norton.

Zablotska, Lydia B. et al. (2008) "A Cohort Study of Thyroid Cancer and Other Thyroid Diseases after the Chernobyl Accident: Dose-Response Analysis of Thyroid Follicular Adenomas Detected during First Screening in Ukraine (1998-2000)." *American Journal of Epidemiology*, volume 167, number 3, February: 305-312. Available at <http://aje.oxfordjournals.org/cgi/content/full/167/3/305>. Accessed March 12, 2008.

Zimmerman, Fred and Dave Beal. (2002) *Manufacturing Works*. New York, NY: Dearborn Trade and Publishing.

# *Glossary*

**Anthracite (coal).** A hard, shiny coal that has a high carbon content. It is valued as a fuel because it burns with a clean flame and intense heat, but without smoke or odor. Also called "hard coal." It is much less abundant than bituminous coal. [See also bituminous coal, lignite.]

**Avoirdupois.** A system of weight (for bodies or goods) based on one pound equaling 16 ounces.

**Back-up power.** The electrical generating power necessary to supplement or "back-up" solar energy and wind power for those periods when the sun is not shining or the wind is not blowing.

**Baseload power.** Baseload power is the power that a utility generates continuously and at a constant rate throughout the year in anticipation of the minimum (base) customer demand (load) that occurs regardless of daily or seasonal fluctuations; the minimum amount of electrical power that a utility can deliver steadily over time.

**Battery Electric Vehicle (BEV).** A vehicle powered by electricity stored in a battery.

**Bituminous (coal).** A soft type of coal that burns with a smoky, yellow flame. Bituminous coal is the most abundant form of coal. It has a high sulfur content, and, when it burns it gives off sulfurous compounds that contribute to air pollution and acid rain. [See also anthracite, lignite.]

**British Thermal Unit (BTU).** A unit of heat equal to the amount of heat energy required to raise the temperature of one pound of water from 60 to 61 degrees Fahrenheit at one atmosphere pressure; equivalent to 251.997 calories.

**Capacity factor.** *See* Service factor

**Carbon dioxide ($CO_2$).** Colorless, odorless, incombustible gas formed during respiration, combustion, and organic decomposition.

**CANDU (CANada Deuterium Uranium).** CANDU is a Canadian-designed nuclear power reactor of PHWR type (pressurized heavy water reactor) that uses heavy water (deuterium oxide) or moderator and coolant, and uses natural uranium for fuel.

**Chlorofluorocarbons (CFCs).** Chemical compounds consisting of carbon, hydrogen, chlorine, and fluorine, once used widely as aerosol propellants and refrigerants. Believed to deplete the atmospheric ozone layer.

**Cooling tower.** Device that dissipates heat from water-cooled systems through a combination of heat and mass transfer, whereby the water to be cooled is distributed in the tower and exposed to circulated ambient air.

**Demand (electric).** Electrical power delivered to a system at a given time or averaged over a designated period. Expressed in kilowatts.

**DUPIC (Direct Use of spent light-water-reactor Plutonium In CANDU reactors).** DUPIC is a manufacturing process that turns an American-style light water reactor's spent fuel pellets into Canadian fuel pellets for use in Canadian reactors (CANDUs). Canadian reactors are much more efficient than light-water reactors and can directly use what is traditionally thought to be spent fuel. There is no change made to the composition of the "spent" fuel. This is not reprocessing; reprocessing *recycles*, but DUPIC *reuses*. Canadian reactors are simply more efficient and can get more energy out of the same amount of enriched uranium.

**Efficiency.** Ratio of power output to input.

**Energy.** The capacity to do work (also known as potential energy) [See Power.]

**Fast Neutron Reactor (FNR).** [also called Fast reactor (FR), Fast-spectrum reactor (FSR), Fast-breeder reactor (FBR), and integrated fast reactor] In a fast neutron reactor, the fission chain reaction is sustained by fast neutrons. On average, more neutrons per fission are produced from fissions caused by fast neutrons than from those caused by thermal neutrons (contrast with "thermal reactor"). A fast neutron reactor can extract energy via fission from all types of uranium, including depleted uranium, which is otherwise waste from enrichment, and from all isotopes of the transuranic elements. Neutrons are kept at higher average energy in a fast neutron reactor than in a thermal reactor, and these higher energy neutrons are able to fission more isotopes. Though conventional thermal reactors also produce excess neutrons, fast reactors can produce enough of them to breed more fuel than they consume. Such designs are known as fast-breeder reactors. One example of such a reactor is the Phénix reactor near Cadarache, France.

**Fossil Fuels.** Energy sources derived from dead plant material. The three main kinds are coal, natural gas, and oil (petroleum).

**GNEP (Global Nuclear Energy Partnership).** The Global Nuclear Energy Partnership (GNEP) seeks to develop a global consensus on expanding the use of economical, carbon-free nuclear energy to meet growing electricity demand. GNEP will use a nuclear fuel cycle that enhances energy security and promotes non-proliferation. The Partnership would achieve its goal by having nations with secure, advanced nuclear capabilities provide fuel services — fresh fuel and recovery of used fuel — to other nations that agree to employ nuclear energy for generating power only. The closed fuel cycle model envisioned by this Partnership requires development and deployment of technologies that enable recycling and consumption of long-lived radioactive waste. In so doing, the Partnership would demonstrate the critical technologies needed to change the way used nuclear fuel is managed

— to build recycling technologies that enhance energy security in a safe and environmentally responsible manner, while simultaneously promoting non-proliferation. See: http://www.gnep.energy.gov/gnepProgram.html

**Gigawatt.** A unit of power equal to one billion watts. [See Watt]

**Gray.** The Systeme International (SI) unit for the energy absorbed from ionizing radiation, equal to one joule per kilogram.

**Joule.** A unit of electrical energy equal to the work done when a current of one ampere passes through a resistance of one ohm for one second

**Kilowatt.** A unit of power equal to 1000 watts. [See Watt]

**Kilowatt hour (kWh).** Unit of electrical consumption equal to the work done by one kilowatt acting for one hour.

**Lignite (coal).** A soft, brownish-black form of coal having more carbon than peat, but less carbon than bituminous coal. Lignite is easy to mine, but it does not burn as well as other forms of coal. It is a greater polluter than bituminous coal because it has a higher sulfur content. [See also anthracite, bituminous coal.]

**Load.** The demand upon the operating resources of a system. In the case of energy loads, the word generally refers to heating, cooling, and electrical (or demand) loads.

**Megawatt.** A unit of power equal to one million watts. [See Watt]

**Millirem.** One-thousandth of a rem. [See Rem]

**Nitrogen (Nitrous) oxides.** Chemical compounds that contain nitrogen and oxygen. They react with volatile organic compounds in the presence of heat and sunlight to form ground-level ozone and are a major precursor to acid rain.

**Octane.** A numerical representation of the anti-knock properties of motor fuel, compared with a standard reference fuel, such as isooctane, which has an octane number of 100. Also called octane rating.

**Off-peak.** A utility rate schedule when the costs of energy and demand are typically less expensive.

**On-peak.** A utility rate schedule when the costs of energy and demand are generally more expensive.

**Organization of Petroleum Exporting Countries (OPEC).** An organization consisting of the world's major oil-exporting nations, OPEC was founded in 1960 to coordinate the petroleum policies of its members and to provide member states with technical and economic aid. OPEC is a cartel that aims to manage the supply of oil in an effort to set the price of oil on the world market in order to avoid fluctuations that might affect the economies of both producing and purchasing countries. Comment: OPEC membership is open to any country that is a substantial exporter of oil and that shares the ideals of the organization. OPEC member nations currently supply about 40 percent of the world's crude oil and 16 percent of its natural gas. At the end of 2003, OPEC nations possessed about 78 percent of the world's total proven crude oil reserves. OPEC's founding members are Iran, Iraq, Kuwait, and Venezuela. OPEC currently has 11 member nations: Algeria, Indonesia, Iran, Iraq, Kuwait, Libya, Nigeria, Qatar, Saudi Arabia, United Arab Emirates (UAE), and Venezuela. (Source of definition and comment: Investopedia, June 2007.)

**Oxide.** Any compound of oxygen with another element or a radical.

**Oxidize, oxidation.** To combine with oxygen, to make an oxide, especially to remove hydrogen. Also, to increase the positive charge or valence of an element (atom or molecule) by removing electrons.

**Peakload power.** The energy generated by utilities to meet energy needs during periods of highest (peak) consumption, as during summer daylight hours when demand for air conditioning is highest.

**pH level.** The measure of acidity or alkalinity in a solution.

**Photocell.** A device that responds electrically to the presence of light.

**Photovoltaics**

1. (used with a singular verb) a field of semiconductor technology involving the direct conversion of electromagnetic radiation, such as sunlight, into electricity.

2. (used with a plural verb) devices designed to perform conversion of electromagnetic radiation into electricity.

**Power.** Work done or energy transferred per unit of time.

**PUREX.** PUREX (or Purex) is the acronym for Plutonium and Uranium Recovery by EXtraction. PUREX is the de facto standard for recovery of uranium and plutonium from nuclear fuel.

**Pyroprocessing.** A generic term for several kinds of pyrometallurgical reprocessing. In a fast neutron reactor this term refers to a process that recycles spent fuel at the reactor site.

**Quad.** One quad equals one quadrillion BTU or about 293 billion kilowatt hours.

**Rad (Radiation Absorbed Dose).** A unit of energy absorbed from ionizing radiation, equal to 100 ergs per gram or 0.01 joules per kilogram of irradiated material. Note: The term rad has been replaced as a standard scientific unit by the gray.

**Rem (Roentgen Equivalent in Man).** A unit for measuring absorbed doses of radiation, equivalent to one roentgen of x-rays or gamma rays; the amount of ionizing radiation required to produce the same biological effect as one rad of high-penetration x-rays. Note 1: a normal medical x-ray delivers about 0.02 rem; a fatal dose of radiation is several thousand rem. Note 2: the term rem has been replaced in most scientific contexts by the term sievert.

**Semiconductor.** Any solid substance, typically crystalline (such as silicon, germanium, and gallium arsenide), that conducts electricity more easily than insulators but less easily than conductors. As crystals of these materials are grown, they are "doped" with traces of other elements called "donors" or "acceptors" to make regions or "holes" in which current will pass in one direction, but not the other, thereby forming a diode. In semiconductors, thermal energy is enough to cause a small number of electrons to escape from the valence bonds between atoms (the valence band). They orbit instead in the higher-energy conduction band, in which they are relatively free. The resulting gaps in the valence band are called holes. Semiconductors (also known as "chips" or "integrated circuits") are used to make diodes, transistors, memory and computer processing circuits, and other "solid state" electronic components and circuitry. (Sources: American Heritage Science Dictionary and Free On-line Dictionary of Computing, accessed July 1, 2007.)

**Service factor [also called capacity factor].** The portion of the time that a facility is operating; the portion of the time that an energy-producing facility is generating energy.

**SI (Systeme International d'Unites or International System of Units).** A comprehensive metric system of measurements used by scientists around the world, with the exceptions of Liberia, Myanmar, and the United States. The fundamental measures include length (meter), mass (kilogram), time (second), electric current (ampere), temperature (kelvin), amount of matter (mole), and luminous intensity (candela).

**Sievert.** The official SI measure of ionizing radiation required to produce the same biological effect as one rad of high-penetration x-rays; equivalent to 100 rem.

**Terrawatt.** A unit of power equal to one trillion watts. [See Watt]

**Ton.** A short ton, also called a freight ton, is an amount weighing 2000 pounds avoirdupois (or 0.907 metric tons). A long ton, a weight unit used in Great Britain, is an amount equal to 2240 pounds (or 1.016 metric tons. A metric ton is a unit of weight equivalent to 1000 kilograms (2204.62 pounds)

**Transesterification.** A reversible chemical reaction between an ester of one alcohol and a second alcohol to form an ester of the second alcohol and an alcohol from the initial ester. For example, the reaction that transforms methyl acetate and ethyl alcohol into ethyl acetate and methyl alcohol.

**Urex Plus (+) or UREX+.** UREX+ is the acronym for URanium EXtraction. This process is a proliferation-resistant modification of the PUREX process.

**Voltage, volts.** International System (IS or SI) unit of electrical potential or the amount of electrical flow, also referred to as "electromotive force."

**Watt.** An International System (IS or SI) unit of power equal to one joule per second.

**Kilowatt (KW).** A unit of power equal to one thousand watts

**Megawatt (MW).** A unit of power equal to one million watts

**Gigawatt (GW).** A unit of power equal to one billion watts

**Terawatt (TW).** A unit of power equal to one trillion watts.